The Evolution of Sex

TO THE MEMBERS OF THE
SEMINAR ON EVOLUTION THEORY
ANN ARBOR, 1976

The Evolution of Sex

JOHN MAYNARD SMITH

Professor of Biology
University of Sussex

CAMBRIDGE UNIVERSITY PRESS

CAMBRIDGE
LONDON · NEW YORK · MELBOURNE

Published by the Syndics of the Cambridge University Press
The Pitt Building, Trumpington Street, Cambridge CB2 1RP
Bentley House, 200 Euston Road, London NW1 2DB
32 East 57th Street, New York, NY 10022, USA
296 Beaconsfield Parade, Middle Park, Melbourne 3206, Australia

First published 1978

Printed in Great Britain
at the Alden Press, Oxford

Library of Congress cataloguing in publication data
Smith, John Maynard, 1920–
The evolution of sex
Bibliography: p. 195
Includes index.
1. Sex (Biology) 2. Evolution 2. Population genetics I. Title.
QH481.S64   574.1'66   77-85689
ISBN 0 521 21887 X
ISBN 0 521 29302 2 paperback

Contents

	Preface	ix
1	The problem	1

2 **Some consequences of sex and recombination – I. The rate of evolution** — 11
Preamble — 11
A Response to selection based on pre-existing genetic variability — 13
B Response to a sudden change in the environment, based on newly arising mutations — 16
C Continuous evolutionary change — 19
D Genetic load, extinction and the Red Queen — 23
E Transient and steady-state models of evolution — 28
F A review of the models — 31

3 **Some consequences of sex and recombination – II. Muller's ratchet** — 33

4 **Could sex be maintained by group selection? The comparative data** — 37
Preamble — 37
A Do parthenogenetic varieties enjoy a twofold advantage? — 38
B The nature, genetic consequences and origins of parthenogenesis in animals — 42
C Parthenogenesis in plants — 49
D The evolutionary potential of parthenogenetic strains — 51
E Cyclical and facultative parthenogenesis – the balance argument — 57
F What is it that goes extinct? — 66
G Conclusions — 69

5   Recombination – the problem   72
   Preamble   72
   A Is there genetic variance for recombination frequency?   73
   B Selection against recombination in a uniform environ-
      ment   77
   C Linkage disequilibrium and recombinational load –
      the observations   83
   D Supergenes and inversions   85

**6   Short-term advantages for sex and recombination – I. An
   unpredictable environment**   89
   Preamble   89
   A Selection in a varying environment   90
   B Spatial variation of the environment   96
   C Sib competition   99
   D Environmental unpredictability – the evidence from
      geographical parthenogenesis   108

**7.   Short-term advantages for sex and recombination – II.
   Selection in a finite population**   111
   Preamble   111
   A Gene selection and hitch-hiking – a digression   112
   B Hitch-hiking and recombination   114
   C Selection for recombination in the presence of recur-
      rent mutations   117
   D Effects of selfing on selection for higher recombination   119
   E Conclusions   122

8   Hermaphroditism, selfing and outcrossing   124
   Preamble   124
   A Selection for self-compatibility in hermaphrodites   125
   B Resource allocation in hermaphrodites   130
   C Other models for hermaphroditism   133
   D Hermaphroditism, monoecy, and dioecy in plants   135
   E Hermaphroditism in animals   138
   F Avoidance of inbreeding in animals   139
   G The human incest taboo   142

9   Anisogamy and the sex ratio   146
   A Methods: evolutionarily stable strategies   146
   B Genetic variance of the sex ratio   148
   C Anisogamy   151

D The sex ratio with random mating   157
E The sex ratio with local mate competition   160
F The sex ratio when the value of a male or female varies
with circumstances   161
G Parent–gamete competition: meiotic drive   163
H Parent–offspring competition   165
Appendix: The evolution of stable sex ratios   166

10 Sexual selection   168
A The concept of 'female choice'   168
B Parental care   175
C Monogamy, polygamy, and sexual dimorphism   183

11 Mutation   188
A Are mutation rates optimal or minimal?   188
B Mutation and hitch-hiking   192
Appendix: The mutation rate in an infinite asexual
population   193

References   195

Author Index   211

Subject Index   215

Preface

The aim of this book is to elucidate the selective forces responsible for the evolution of sex, of recombination rates, of breeding systems, and of mutation rates. I might have called it 'The Evolution of Genetic Systems' were it not that a classic with that title already exists. But whereas Darlington approached these topics from the standpoint of a cytologist with a distrust of mathematical reasoning, my own approach is that of a population geneticist.

It will be obvious that I have been greatly influenced by G. C. Williams's *Sex and Evolution*. I share with him a distaste for the Panglossian belief that if some characteristic can be seen as benefiting the species, then all is explained. I am under no illusion that I have solved all the problems which I raise. Indeed, on the most fundamental questions – the nature of the forces responsible for the maintenance of sexual reproduction and genetic recombination – my mind is not made up. On sex, the relative importance of group and individual selection is not easy to decide. On recombination, group selection can hardly play a significant role, but it is not clear to me whether the short-term selective forces I discuss are sufficient to account for the facts, or whether models of a qualitatively different kind are needed.

Inevitably, this uncertainty will make the book harder to follow. An author who knows his own mind about everything can present a clear and consistent case. I have felt more that I was carrying on a debate with myself, presenting the arguments first on one side and then on the other. To help the reader, I have provided a 'preamble' to most of the chapters, and I have not been afraid to repeat myself if it seemed to make for clarity.

I have made no attempt to present an exhaustive review of the comparative data on breeding systems. Instead, I have tried to put the theoretical issues as clearly as I can, and to give enough of the evidence to show what *kinds* of facts might be relevant. It has struck me, while writing, that the crucial evidence is often missing, simply because the theoretical issues have not been clearly stated, so that the relevance of a particular fact has not been appreciated. If I do no more than encourage experimentalists and field workers to collect the relevant data, I shall be well satisfied.

There are several aspects of the evolution of genetic systems about which I say little. These include chromosome structure, the genetics of sex determination and self-incompatibility, the significance of haplo-diploid life cycles, and parasexual processes in prokaryotes. I have made no attempt to discuss the molecular basis of recombination and mutation. All these topics are relevant to my general theme, but I have preferred to stick to topics about which I have something new to say.

The major part of the writing was done while I was a visitor at the Museum of Zoology at the University of Michigan, Ann Arbor. While there, I ran a seminar at which many of the topics in the book were discussed. I have exploited the graduate students who looked up references, told me of their own work, and suggested ideas; my only excuse is that I warned them that I would do so. I am grateful to them for making my stay at Ann Arbor so stimulating. The book has been read in manuscript by my colleagues Brian and Deborah Charlesworth, by Joe Felsenstein of the University of Washington, and by Eric Charnov and Jim Bull of the University of Utah. I have been helped greatly by their comments, although I have not always taken their advice.

Sussex University   J. Maynard Smith
July 1977

1 *The problem*

This chapter should be read as a guide to the rest of the book. In it, I outline what seem to me to be the central problems to be solved, I indicate briefly the theories and kinds of evidence which may be relevant to their solution, and, when it seems helpful, direct the reader to the chapters and sections in which each matter is discussed further.

The central question is as follows: what selective forces maintain sexual reproduction and genetic recombination in nature? It may help to classify the various theories; first, according to the time scale on which selection is supposed to act, and then according to the 'unit of selection' – population, individual, or gene. Classifying according to time scale, we have at one extreme selective forces comparable to those which established industrial melanism or insecticide resistance in natural populations. Such a selective force could, at least in principle, increase the frequency of an allele for high recombination relative to its low-recombination allele in each generation. Such forces exist (Chapter 6), but it is far from clear that they are the only or the most important ones.

At the other extreme are forces which produce an increase in recombination only when their effects are summed over a period long enough for some populations to go extinct and for new ones to arise. I return to such forces in a moment. But there are, in addition, selective forces acting within a population, and over an intermediate time scale. There is no guarantee that the frequency of an allele for high recombination will increase in any one generation, but it will do so when summed over tens or hundreds of generations (Chapter 7). Such processes are hard to treat analytically, but they may well be the most important ones with which we are concerned.

A more usual way of classifying selective forces is in terms of 'group' and 'individual' selection. Thus, natural selection will operate on any set of entities with the properties of multiplication, heredity, and variation. If the entities in question are individuals, we can speak of 'individual selection'; it is this type of selection with which the vast majority of evolution theory is concerned. If the entities are groups – species, or reproductively isolated populations – we speak of 'group selection'. It was Fisher's view that sexual reproduction was the only adaptation which had evolved by group selection, but he at least hinted that recombination might be explained without group selection (see p. 119).

The reason why most evolutionists have been reluctant to accept group selection as a significant force in evolution is that the circumstances in which a trait advantageous to the group but deleterious to the individual could increase in frequency are restricted and unlikely to arise often in nature (Maynard Smith, 1976c; for a contrary view see Gilpin, 1975). My own active interest in the evolution of genetic systems dates from around 1964, when I recognised that it was illogical to reject Wynne-Edwards's (1962) views in ecology as group selectionist, if at the same time I accepted a group selectionist explanation of sex. G. C. Williams's interest in genetic systems seems to have been aroused in a similar way, although he appears to have been provoked more by Emerson (1960) than by Wynne-Edwards.

Is it then possible that sex is maintained by group selection? In its simplest form, the argument would go as follows. An individual female which abandons sexual reproduction obtains thereby an immediate short-term advantage. This advantage means that once a parthenogenetic strain is established, it will selectively displace the sexual species. In the long run, however, the new parthenogenetic species is doomed to extinction because of its inability to evolve. While parthenogenetic species go extinct, sexual ones speciate, thus maintaining the total number of species. The result of this process is that, at any one time, most species are sexual.

Thus, the group selection explanation of sexual reproduction accepts that there is a short-term advantage to parthenogenesis, and argues that this is more than balanced by the long-term advantages of sex. I now discuss these two points.

Consider first the twofold disadvantage of producing males. Although absurdly simple, the point is so fundamental and so often misunderstood that I repeat below the argument first presented by Maynard Smith (1971*b*). Suppose that, in a sexual species, with equal numbers of males and females, a mutation occurs causing females to produce only parthenogenetic females like themselves. The number of eggs laid by a female, k, will not normally depend on whether she is parthenogenetic or not, but only on how much food she can accumulate over and above that needed to maintain herself. Similarly, the probability, S, that an egg will survive to breed will not normally depend on whether it is parthenogenetic. With these assumptions the following changes occur in one generation:

	Adults		Eggs	Adults in next generation
Parthenogenetic ♀♀	n	⟶	kn	⟶ Skn
Sexual { ♀♀	N	⟶	$\frac{1}{2}kN$	⟶ $\frac{1}{2}SkN$
♂♂	N	⟶	$\frac{1}{2}kN$	⟶ $\frac{1}{2}SkN$

Hence, in one generation, the proportion of parthenogenetic females increases from $n/(2N+n)$ to $n/(N+n)$; when n is small, this is a doubling in each generation.

Some of the confusion which has arisen over this came, I think, from the phrase 'the cost of meiosis'. In a species with isogametes there is no necessary twofold cost associated with meiosis, although the time taken to complete the meiotic divisions might constitute a cost. In a sense, a gene in a primary oocyte is running a 50% chance of being eliminated in a polar body, and could therefore gain a twofold selective advantage by suppressing meiosis. But I believe that the twofold advantage of parthenogenesis is best seen as the advantage of not producing males. Some situations in which the advantage does not accrue are discussed in Chapter 4, section A.

This short-term advantage of parthenogenesis has been noted in passing by a number of authors. However, it has not led them to reformulate the question of the evolution of parthenogenesis in what seems to me to be the natural way: i.e. why do parthenogenetic varieties not replace sexual ones? For most writers, the question seems to have remained, why are some parthenogenetic varieties able to establish themselves? There are, I think, two contributing reasons why the question is still put in this way. First, some authors have thought that a twofold increase in reproductive rate would be an advantage only when a population is increasing in numbers, a confusion which has not been helped by all the recent discussion of r and K selection and its relation to colonisation.

A second reason why the twofold advantage has, until recently, been either ignored or treated as unimportant is that most writers on the subject have felt that the short-term advantages of sexual reproduction in an unpredictable environment are so overwhelming that the problem is to explain why parthenogenesis exists, not to explain why it is not universal. I hope that Chapter 6 will dispel any idea that selection in an unpredictable environment necessarily or easily confers an advantage on sex. I am, however, very conscious of the fact that my method of formulating the problem of selection in an unpredictable environment may have caused me to draw misleading conclusions. But if anyone believes that, let him show how the problem should be expressed.

The second point to be established is that there are long-term selective advantages in sexual reproduction to the species. The nature and extent of these advantages are discussed in Chapters 2 and 3. The essential point, however, was made independently by Fisher (1930) and Muller (1932). In a population without genetic recombination, if two different favourable mutations, $a \rightarrow A$ and $b \rightarrow B$, occur in separate individuals, there is no way in which they can be combined in a single descendant. An AB individual can arise only if a second A mutation occurs in a descendant of the original B mutant (or a second B in a descendant of the original A). With recombination, an AB individual can arise by recombination in a common descendant of the two original mutants.

Thus the essential logical requirements for a group selection explanation exist, namely genetically isolated groups (sexual species, and parthenogenetic clones), and a trait which is individually disadvantageous but advantageous to the survival of the group. The only questions that remain are, first, is the group selection hypothesis plausible in numerical terms, and, secondly, is it supported by the comparative data?

The question of numerical plausibility is easily formulated. The group selection hypothesis in its simple form requires that for every successfully established parthenogenetic clone, the sexual species which gave rise to it must go extinct. Since species extinction is not an everyday occurrence, this requires that parthenogenetic origins must likewise be rare. There is no easy way to estimate how frequently such origins occur. The problem is discussed in Chapter 4, sections B and C. The situation is somewhat different between plants and animals because most spontaneous parthenogenesis in animals is of a kind which gives rise to individuals of low viability because of genetic homozygosity. At least in animals it is not implausible to suppose that viable parthenogenetic varieties may arise sufficiently rarely.

Further support for the group selection argument comes from the arguments in Chapter 4, sections D and F. Section D is primarily concerned with showing that the taxonomic distribution of parthenogenetic varieties supports the hypothesis that such varieties are doomed to early extinction, presumably because of lack of evolutionary potential. Section F points out that the establishment of a successful parthenogenetic variety is often not followed by the extinction of the sexual species which gave rise to it, because the variety may be adapted to only a part of the ecological range of the parent species.

All this, however, merely points to the plausibility of the group selection hypothesis; it does not prove it to be true. To my mind, much the most powerful argument of an empirical kind against the hypothesis was proposed by Williams (1975); it is the 'balance' argument. Suppose that a population consists of facultatively parthenogenetic organisms which can reproduce either sexually or asexually. In such a case it is inconceivable that there should not be genetic variation capable of wholly suppressing sex if this were advantageous in the short run. Since sex continues, it must have some short-term advantages.

The argument is a powerful one. It is discussed in Chapter 4, section E. The difficulty is not with the argument but with the data. There are some unequivocal cases which support Williams's position. But, in assessing the data, it is essential to remember that a population consisting of a mixture of sexually reproducing individuals and obligate parthenogenetic ones does not support the balance argument because it is possible that the parthenogens contain a different (and narrower) range of genotypes. Further, a species in which individuals produce sexual and asexual eggs with different ecological roles (as, for example, the overwintering sexual and immediately hatching asexual eggs of rotifers, or the sexual dispersive seeds and asexual bulbils of onions) does not support the balance argument; it could be that sex is retained in these cases only because of its ecological correlates. If, in such cases, it can be shown that there are genetic variants capable of producing all ecological types asexually, this would support the balance argument.

I am sure that there must be much data on the balance argument with which I am unfamiliar, and still more which could easily be collected if its relevance were appreciated. In animals, in particular, there are many reports of populations in nature with a substantial excess of females. But are these facultative parthenogens, or are they mixtures of sexual individuals and obligate parthenogens?

I do not find it possible to give an unequivocal answer concerning the role of group selection in the maintenance of sexual reproduction. It has played some role, as evidenced by the taxonomic distribution of parthenogens; but it is not the only relevant force, as will be apparent from the review of the balance argument in Chapter 4, section E. But, whatever one may think of the role of group selection in the maintenance of sex, it cannot explain how it started, and it cannot explain the maintenance of high levels of genetic recombination within sexual populations.

So far I have discussed only the maintenance of sexual reproduction and genetic recombination. But what of their origins? Surely evolution theory should be concerned with the origins of adaptations, not merely with their maintenance once they have arisen. There is much in this objection; indeed one of the things I have learnt while writing this book is that my own insight into the field may have been obscured by an obsession, which I share with most population biologists, with equilibrium situations.

Yet there is really little alternative. Recombination probably originated some three thousand million years ago, and eukaryotic sex one thousand million years ago. Each origin may have been a unique series of events. We can speculate about such events, but cannot test our speculations. In contrast, selection must be acting today to maintain sex and recombination. We have to concentrate on maintenance rather than origins because only thus can we have any hope of testing our ideas. Nevertheless, something must be said about origins.

Although group extinction may eliminate the occasional 'loss' mutation, it cannot have brought together the series of adaptations concerned in the origin of sex. Long before the origin of eukaryotic sex, the prokaryotes had acquired the capacity for genetic recombination – that is, the pairing, breaking, and rejoining of homologous lengths of DNA. It seems clear that its original function was not the generation of evolutionary novelty, but the repair of damage. This point is discussed further on p. 36.

In the prokaryotes, the problem of bringing together in a single cell DNA from two different ancestors is solved in a variety of ways. But with the evolution of the eukaryotic cell, the problem had to be solved anew. The origin of meiosis, syngamy, and the haplo-diploid cell cycle remains one of the great puzzles of evolution theory. The best scheme I can offer is illustrated in Figure 1. Starting from a population of haploid single-celled organisms, one can at least guess at the main stages and the selective forces responsible. The first stage would be binary cell fusion to form a heterokaryotic cell, with two haploid nuclei of different ancestry. The selective advantage of such fusion would be analogous to the advantages of hybrid vigour, particularly the covering up, by complementation, of deleterious genes. As in the origin of recombination, the first step was to compensate for damage rather than to create novelty.

To evolve from a heterokaryon to a diploid it is required only that the two nuclei use a single spindle in mitosis rather than each its own spindle. This would have the obvious advantage of ensuring that one copy of each nuclear chromosome set was transmitted to each daughter cell, retaining the advantages of hybrid vigour; if two spindles were used, there would always be danger of two diploid 'homozygotes' resulting from mitosis.

So far, all is straightforward. But given the evolution of a successfully replicating diploid, why then evolve the pairing of homologous chromosomes, chiasma formation, reduction divisions

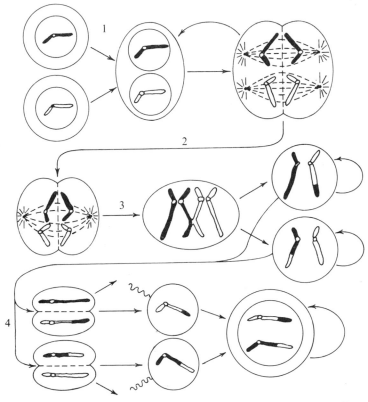

Figure 1. A scheme for the origin of sex. 1, Fusion of haploid cells to form heterokaryon. 2, Origin of diploid from heterokaryon. 3, Chiasma formation during mitosis ('somatic crossing over') generates variability within a clone. 4, Reduction and syngamy restores heterozygosity.

and subsequent syngamy? It is possible to suggest why it might be selectively advantageous to evolve homologous pairing at the 'four-strand' stage, accompanied by chiasma formation (the genes for which would already be present) between non-sister strands. The consequence of such 'somatic crossing over' is the generation of new genetic variants within a clone. But somatic crossing over generates new variation at the expense of genetic homozygosity. At this stage, a final reduction division followed by syngamy would have been favoured because it again restored the lost genetic heterozygosity.

There remains the question of why, given a haplo-diploid cycle, both higher animals and plants spend the major part of the cycle as diploids. We do not know. Two possibilities are that diploidy protects against the deteriorative effects of somatic mutation, and that it in some way assists complex processes of cellular differentiation, but these are unsupported speculations.

I cannot pretend to much confidence in this explanation. Apart from other drawbacks, it argues in a circle: first, give up genetic variability for hybrid vigour; then abandon hybrid vigour for the benefits of variability; finally, regain hybrid vigour through reduction division and syngamy. It also runs contrary to the orthodox view that in primitive sexual eukaryotes the main part of the life cycle was spent as a haploid. But at least it may provoke better speculations.

More important than the failure of group selection to explain the origin of sex is its failure to explain the maintenance of genetic recombination. What evidence we have (reviewed in Chapter 5, section A) suggests that there is within-population genetic variation for the frequency and location of chiasmata. If it were advantageous in the short run to reduce or eliminate crossing over, it seems that the genetic variation exists upon which selection could act. True, there is no twofold disadvantage to recombination comparable to that associated with sex. Yet there almost certainly is a short-term disadvantage to recombination. In Chapter 5, the theoretical reasons for this are reviewed in section B, and some observational evidence supporting the theory in section C. In effect, what both theory and observation suggest is that it is selectively disadvantageous to break up the co-adapted genotypes of surviving adults by recombination.

Thus Williams's (1975) balance argument applies to the maintenance of recombination, whatever the difficulties may be of applying it to the maintenance of sex. Therefore, unless we are going to claim that chiasmata perform some essential mechanical role in meiosis – and it would be difficult to support such a claim in view of the absence of chiasmata in the meiosis of male Diptera – we are driven to seek some counterbalancing short-term advantage for recombination.

The search for such short-term advantages is taken up in Chapters 6 and 7, which form the core of the book. There are basically three types of model: 'unpredictable environment'; 'sib competition'; and the 'Hill–Robertson effect' i.e. the effects of randomly generated linkage disequilibrium. Felsenstein & Yokoyama (1976) would reduce these to two by treating the 'sib-competition' model as a sub-class of the Hill–Robertson effect.

Each type of model has its difficulties. If my analysis is sound, an unpredictable environment will favour increased recombination only when the type of unpredictability is of a special and, in my view, implausible kind. Sib competition, which I believe to be the essential new feature of the models proposed by Williams (1975), can favour higher recombination in an environment which is unpredictable in a more plausible way. The major snag is that, as soon as one allows for the fact that alleles at a number of different loci may be concerned with adaptation to a single feature of the environment, the direction of selection is reversed and favours tighter linkage.

The Hill–Robertson effect is sufficiently universal to be the type of mechanism we are seeking. It arises in two forms:

(i) the 'hitch-hiking' effect (Chapter 7), in which recombination favours the establishment of new favourable alleles; and

(ii) 'Muller's ratchet' (Chapters 2 and 7, section C), in which recombination favours the elimination of deleterious alleles.

In both cases, it can be shown that there is short-term selection for higher recombination.

I fear that the reader may find these models insubstantial and unsatisfactory. But they are the best we have.

The topics discussed in Chapters 8–10 are in a sense subsidiary to the main theme. In these chapters, I take it for granted that sexual reproduction exists, and discuss the selective factors responsible for variations in the means of achieving it. Finally, in Chapter 11 I take up a new topic – the evolution of the mutation rate. The major part of the book is concerned with the evolution of those genes which determine the rate of recombination between other genes; Chapter 11 is concerned with the evolution of those genes which determine the accuracy of replication of other genes.

2 Some consequences of sex and recombination – I. The rate of evolution

Preamble

It is an essential part of the group selection argument concerning the maintenance of sex that populations with sex and recombination can evolve more rapidly than those without. In this chapter I discuss how far and in what circumstances this is true. The answers to these questions are important, whether or not the group selection argument is correct.

When a sexual population is exposed to new selection pressures, for example under domestication, or in the laboratory, or in nature when there is a sudden change in environmental conditions, there is usually rapid and adaptive response in the genetic constitution of the population. If directional selection for some phenotypic character is maintained in the laboratory for a number of generations, the adaptive response will slow and come to a halt. A classic example is shown in Figure 2. When the 'plateau' has been reached there is no longer any appreciable additive genetic variance for the selected trait, although there may be considerable non-additive genetic variance (Clayton & Robertson, 1957). Further progress then depends on new mutation, or on the release of genetic variability by recombination. The latter possibility arises because, after intense selection in a relatively small population, the selected population may, for two linked loci, contain only $+ -$ and $- +$ chromosomes.

The extent and rate of the response depends on the nature of the genetic variability present in the foundation population. However, it is quite usual to be able to produce a rate of change in some metric character, greater than 1% of the mean value of that character, in each generation for 20–30 generations. This rate of change contrasts very strikingly with the rate with which metric characters change in evolution. For example, during most of the Tertiary Period, the rate of change in the shape of the molar teeth of horses (height/width) was less than 1% per million years, accelerating to 5% per million years during the period of rapid change in the horse lineage which changed from browsing to grazing during the Miocene.

There are, of course, a number of reasons for this difference, of at least five orders of magnitude, between rates of change in evolution and those in artificial selection experiments. In nature, more than one

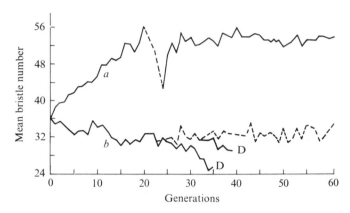

Figure 2. Some results of an experiment by Mather & Harrison (1949) in which, starting from a population of flies with an average of thirty-six abdominal bristles, lines were selected for increased *(a)* and decreased *(b)* bristle number. The dashed lines indicate populations in which selection was relaxed. D indicates populations which died out owing to infertility.

phenotypic trait is exposed simultaneously to selection. The total intensity of selection in domestication can often be greater, because accidental deaths are fewer. Selection in nature may be intermittent, or may reverse its direction. But I think that the most important reason is that the response to artificial selection depends on genetic variability already present, and soon exhausts that variability, or at least its additive component, whereas in evolution there must be, in the long run, a balance between the exhaustion of variability by directional selection and its generation by mutation.

The nature of this balance between selection and mutation is discussed further on pp. 28–31. For the present, however, the important point is that, when comparing rates of evolutionary response, we must distinguish between responses based on pre-existing genetic variability, and responses based on genetic variability arising by mutation after selection has been applied. In fact there are three model situations which can be considered:

(a) response to selection based on pre-existing genetic variability;
(b) response to a sudden environmental change based on new mutations; and
(c) continuous evolutionary change.

A Response to selection based on pre-existing genetic variability

We suppose that at some time, $t = 0$, there is a sudden change in the environment and hence in the relative fitnesses of genotypes. We then compare the rate of response of populations with and without sexual reproduction. The same analysis applies to sexual populations with and without recombination between the selected loci.

Clearly, a sexual population containing a number of different genotypes will respond to selection, whereas a single asexual clone will not. We, therefore, compare the response of a sexual population with that of a population consisting of a set of genetically different asexual clones. For simplicity, I shall consider haploid populations, but the conclusions drawn will hold qualitatively for diploids.

If fitness differences are confined to alleles at a single locus, then there will be no difference between the rates of response of sexual and asexual populations. As Fisher (1930) remarked, 'The only groups in which we should expect sexual reproduction never to have developed would be those, if such exist, of so simple a character that their genetic constitution consisted of a single gene.'

Thus the simplest relevant model is one with two alleles at each of two loci, as follows:

Genotype	ab	Ab	aB	AB
Fitness	1	V	W	kVW
Frequency	p_{ab}	p_{Ab}	p_{aB}	p_{AB}

Consider an asexual population with discrete generations in which p_{ab} represents the frequency of the ab clone in one generation, and p'_{ab} the frequency in the next generation. Then

$$p'_{ab} = p_{ab}/T,$$
$$p'_{Ab} = p_{Ab}V/T,$$
$$p'_{aB} = p_{aB} W/T,$$
$$p'_{AB} = p_{AB} kVW/T,$$

where $T = p_{ab} + p_{Ab}V + p_{aB}W + p_{AB}kVW.$ (2.1)

Consider first the case with fitnesses multiplicative, i.e. with $k = 1$. This is what we expect if selection acts independently on the two loci; for example, if it acts on the A and B loci at different stages in the life history.

Suppose that in one generation the genotype frequencies are in 'linkage equilibrium'; that is, the alleles at the two loci assort independently, so that $p_{ab} \times p_{AB} = p_{Ab} \times p_{aB}$. Then, with $k = 1$, it follows at once from equations (2.1) that $p'_{ab} \times p'_{AB} = p'_{Ab} \times p'_{aB}$; that is, that the frequencies are in linkage equilibrium in the next generation, and will remain so indefinitely.

This conclusion has been drawn for an asexual population. The only effect which sexual reproduction and free recombination can have on genotype frequencies is to bring them closer to linkage equilibrium. Thus the 'linkage disequilibrium', $D = p_{ab} \times p_{AB} - p_{Ab} \times p_{aB}$, is halved by one generation of free recombination. But if D is initially zero, the sexual and asexual populations behave identically.

Thus, we have proved that for multiplicative fitnesses, and genotype frequencies initially in linkage equilibrium, sexual reproduction and recombination make no difference to the rate of evolution (Maynard Smith, 1968a). The conclusion holds for multiple alleles, and for n loci.

If we relax the assumption about multiplicative fitnesses, this may favour either sexual or asexual reproduction (Eshel & Feldman, 1970). Thus suppose that $V, W > 1$, so that evolution is proceeding from $ab \rightarrow AB$. Then if $k > 1$, the asexual population will evolve more rapidly (as measured by the rate of increase of the optimal genotype AB), whereas if $k < 1$, the sexual population will evolve more rapidly.

These conclusions are fairly easy to see intuitively. Thus with $k > 1$, AB genotypes increase rapidly in the absence of recombination, but are broken up by recombination more rapidly than they are synthesised. Indeed, there are some evolutionary changes which can occur in an asexual population but not in a sexual one. For example, if V, $W < 1$ but $kVW > 1$, the evolutionary adaptation $ab \rightarrow AB$ can occur only in the absence of sex (or, with sex, by drift in small populations).

To understand the general evolutionary advantage conferred by sex, we have to consider the assumption concerning linkage equilibrium. We again consider the case in which AB is the fittest genotype. Clearly, if p_{AB} is initially greater than expected from the frequencies of A and B separately, recombination will slow down evolution, whereas if p_{AB} is initially lower than expected, recombination will accelerate evolution (Karlin, 1973). Are there general reasons why we would expect p_{AB} to be lower than its expectation on the assumption of linkage equilibrium?

I suggested one such reason (Maynard Smith, 1968a). A population will often be exposed to a new environment, not because conditions change locally, but because it colonises a new region. Suppose that in the new region the optimal genotype is AB, and that colonists enter this region not from one source but from two, in one of which the optimal genotype was Ab and in the other aB. The new population will then consist mainly of Ab and aB genotypes, and recombination will greatly accelerate evolution.

A second, and possibly more fundamental, reason has been emphasised by Felsenstein (1974). Actual populations are finite and not infinite. Linkage disequilibrium is bound to arise by chance in a finite population. The relevance of this fact will first be discussed for the particular two-locus model now under consideration; its more general relevance is discussed in Chapter 7.

Consider first an infinite asexual population. We suppose that, prior to the environmental change at time $t = 0$, genotypes Ab, aB, and AB are rare, and that their fitnesses are $1 - s_A$, $1 - s_B$ and $(1 - s_A)(1 - s_B)$, respectively. The rate of mutation from $a \rightarrow A$ is μ_A and from $b \rightarrow B$ is μ_B, where $\mu_A \ll s_A$ and $\mu_B \ll s_B$. At equilibrium between mutation and selection, the frequencies of genes A and B are

μ_A/s_A and μ_B/s_B, respectively, and the genotype frequencies are in linkage equilibrium:

$$
\begin{array}{cccc}
ab & Ab & aB & AB \\
\left(1-\dfrac{\mu_A}{s_A}\right)\left(1-\dfrac{\mu_B}{s_B}\right) & \dfrac{\mu_A}{s_A}\left(1-\dfrac{\mu_B}{s_B}\right) & \left(1-\dfrac{\mu_A}{s_A}\right)\dfrac{\mu_B}{s_B} & \dfrac{\mu_A}{s_A}\dfrac{\mu_B}{s_B}
\end{array}
$$

Thus, in an infinite population without recombination, the genotype frequencies reach linkage equilibrium under mutation and selection alone, provided fitnesses are multiplicative. But what of a finite population? Clearly the *expected* frequencies are in linkage equilibrium. But the expected frequency of genotype AB is of the order of the square of the mutation rate. Most finite populations will lack genotype AB altogether. Thus most finite populations will be in linkage disequilibrium of the type in which recombination would be advantageous. As the frequencies of genotypes Ab and aB increased under selection (after the change of environment), AB genotypes would be produced by recombination in a sexual population, whereas in an asexual one they could arise only by mutation.

As the number of loci at which selection is acting increases, the importance of linkage disequilibrium arising by chance in finite populations will increase. If there are many loci, some genotypes are certain to be absent, even if the frequency of particular alleles is not very low.

To summarise, recombination may accelerate evolution if the initial population is out of linkage equilibrium. Linkage disequilibrium is likely to arise in asexual populations because of spatial heterogeneity in the environment, and because of chance events in finite populations.

B   Response to a sudden change in the environment, based on newly arising mutations

Suppose that at time $t=0$ there is a sudden change in the environment, so that at a number L of different loci the existing alleles, say $a, b, c, \ldots$ cease to be optimal. Contrary to the assumption made in the previous section, we now suppose that the favourable alleles

$A, B, C \ldots$ are initially absent. We ask, how much more rapidly will a sexual population adapt, in the sense of reaching, say, an average frequency of 0.95 of the favourable alleles?

This question was considered by Maynard Smith (1971a), using a mixture of analytical and simulation techniques. In brief, the conclusion was that for small populations, sex would confer no advantage; for large populations (roughly, $N > 1/\mu$, where N is the population size and μ the mutation rate per locus) the asexual population would take approximately L times as long to complete its adaptation. This agrees with Fisher's (1930) conclusion that 'The comparative rates of progress of sexual and asexual groups occupying the same place in nature, and at the moment equally adapted to that place, are therefore dependent upon the number of different loci in these sexual species, the genes in which are freely interchangeable in the course of descent', although Fisher did not make any qualification about population size, and, characteristically, did not explain how he reached his conclusion.

This conclusion can be justified intuitively (Figure 3). Consider first a haploid sexual population of size N. After g generations, the expected number of A mutants which has occurred is $Ng\mu$, and of these a fraction of approximately $2s$ will be established in the population, where s is the selective advantage per locus, the rest suffering chance elimination (Moran, 1962). Hence, if g_e is the expected number of generations before an A mutant is established, $Ng_e\mu \times 2s = 1$, or

$$g_e = 1/2 N\mu s. \tag{2.2}$$

If N is large (Figure 3a), g_e will be small compared to g_f (the number of generations elapsing from the establishment of a favourable mutant A to the time when it reaches a frequency of 0.95). Since in a sexual population each favourable mutant increases in frequency more or less independently of the others, the time taken for all L loci to reach 0.95 is approximately g_f generations.

Now consider a large asexual population (Figure 3b). Once the first favourable mutant, say A, has been established, a second favourable mutant contributes to the adaptation of the population only if it occurs in a direct descendant of A. The expected time, g_s say, between

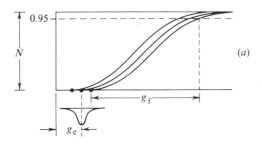

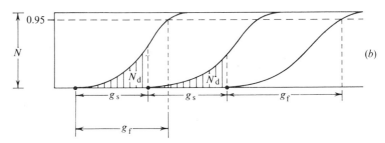

Figure 3. Fixation of L favourable mutations ($L = 3$) in a sexual (a) and an asexual (b) population, when the population size, N, is large. For other symbols, see text.

the establishment of the first favourable mutant, and of a second favourable mutant among the descendants of the first, will be such that $N_d\mu(L-1) \times 2s = 1$, where N_d is the number of direct descendants of the original A mutant who have existed. Provided that L is not very large, N_d must be large. Thus g_s must be large. In fact, g_s will not be as great as g_f (the time to reach a frequency of 0.95), but will not be very much less than g_f, since most of the time to the 0.95 frequency elapses while the gene is still rare. It follows that the time for the asexual population to complete its adaptation is approximately Lg_f, or L times as long as the sexual one. To the extent that $g_s < g_f$, this comparison will exaggerate the advantage of sex. But we are interested only in an approximate answer.

Consider now a small population, for which $g_f \ll g_s$ (Figure 4). Then each favourable mutation goes to fixation before the next occurs. Only one allele difference is segregating in the population at any one time, so that recombination cannot affect the rate of evolution. This point was made by Muller (1932), and has been generally accepted since. Only Bodmer (1970) has queried it; the reason why he reached a different conclusion is that he asked a different question, namely, given that two rare favourable mutations exist in a population, how long will it be before the first individual occurs carrying both favourable alleles?

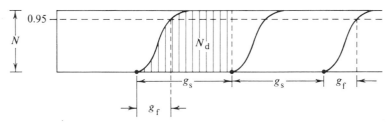

Figure 4. Fixation of L favourable mutations ($L = 3$) in a small asexual population. Each favourable mutation is fixed before the next occurs. For other symbols, see text.

C   Continuous evolutionary change

Suppose now that the environment changes continuously, and that the population tracks it continuously by genetic change. How do sexual and asexual populations compare in this situation? The first serious attempt to answer this question quantitatively was by Crow & Kimura (1965). Their answer is instructive, though I believe it to be misleading. They consider first the response of an asexual population (Figure 5 *top*). Let N be the population size, U the rate per genome at which favourable mutations occur, and s the selective advantage per favourable mutant. The rate of evolution can then be measured by G, the expected number of generations between the establishment in the population of one favourable mutation, say A, and the establishment of a second favourable mutant, say B, among the descendants of A.

G can be calculated from the fact that the expected number, N_G, of descendants of A which must exist before a mutant B is established can be estimated from the equation $N_G U = 1$. (Strictly, this overlooks the fact that not all favourable mutants which occur are established. Kimura & Ohta (1971) allow partially for this, but since this is not my main objection to the Crow & Kimura model, I will not discuss this

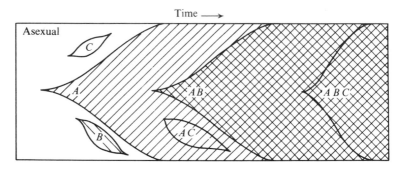

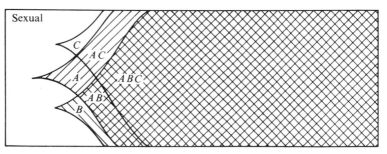

Figure 5. Evolution in sexual and asexual populations. (From Crow & Kimura, 1965, after Muller, 1932.)

point further here: the full correction for this effect is discussed in Maynard Smith, 1971a.)

Given that $N_G U = 1$, it is possible to show that $G \simeq \log N/s$. Hence, the rate of evolution of an asexual population is measured by the fact that one new favourable mutation, on average, is incorporated every G generations.

What of a sexual population? It is near enough true that every favourable mutation which occurs (and which survives the initial chance of accidental elimination) can be established (Figure 5, *bottom*). In G generations, the expected number of new favourable mutants that occur is $NGU = NU \log (N/s)$. Thus while the asexual population incorporates one new mutant, the sexual one incorporates $1 + NU \log (N/s)$ new mutants. Thus the superiority of the sexual population is greatest in large populations when the total rate of favourable mutations is high, and the selective advantage per mutant is small.

This argument holds only if every favourable mutant is different from every other one. Thus in the limit the Crow & Kimura model implies that in an infinite population, the ratio of the rate of evolution of sexual to that of asexual populations would be infinite. Clearly, this cannot be the case. The fallacy arises for the following reason; the number of favourable mutants occuring in G generations is indeed NGU, but if this number is large, many of them will be identical.

It follows that $1 + NU \log (N/s)$ is an overestimate of the advantages of sex. Unfortunately, it is easier to point out the fallacy than to correct it. I think we have to attack the problem from a somewhat different angle. The question we are interested in is this: can a population survive in a changing environment? If the environment changes in a sufficiently radical way, the population will go extinct whatever its genetic mechanism; this might be described as 'running out of niche' (see p. 52). But there will be situations in which a population with genetic recombination can track an environmental change and survive, when an asexual population cannot. So our question becomes: How rapidly can the environment change without causing extinction?

The implication of this question is that, given time, the population could adapt. Because the environment is changing, however, the actual adaptation of a population at any instant is imperfect. We can express this by defining the mean fitness of the population at any instant as $\bar{W}$, and the fitness the population could achieve if conditions were held constant until it had completed its adaptation as W_{max}. Thus $\bar{W}$ is the mean fitness of the contemporary population and W_{max} is the fitness of an individual carrying all alleles favourable in the contemporary environment, whether they are already segregating in the population or not.

$\text{Log} (W_{max}/\bar{W})$ is then a measure of the intensity of directional selection acting on the population. If the environment stays constant, the intensity of selection will in time fall to zero. Similarly, $(W_{max} - \bar{W})/W_{max}$ is a measure of the 'genetic load' arising because the population is not optimally adapted to the environment. I have called this the 'lag load' (Maynard Smith, 1976*b*). It measures the genetic load arising because the population has lagged behind a changing environment. (An argument essentially similar to the one based on the 'lag load' was proposed by Felsenstein (1971); if there is any value in this way of looking at evolution, the credit is his and not mine.)

The lag load differs from the more usual 'substitutional load' in its definition of W_{max}. For the substitutional load, W_{max} includes only favourable mutants already segregating in the population, whereas for the lag load, favourable mutants are included even if they are only potential. The use of the lag load has a number of advantages. First, it removes the apparent absurdity associated with the substitutional load, which increases when a favourable mutation occurs; the lag load does not alter when a favourable mutation occurs in an individual and is reduced when that mutant is fixed in the population. The second advantage is that it provides a measure of the extent to which a population has failed to adapt to the contemporary environment.

A third advantage of the lag load becomes apparent when we attempt to compare the rates of evolution of populations with different genetic mechanisms. We picture evolutionary change as a kind of 'error regulation' in the cybernetic sense. Thus the population tracks the environment by changing genetically. At any instant, the 'error' is measured by the lag load, and the rate of evolution increases as the lag load increases. The more rapid the environmental change the greater is the lag load, and the more rapid is the genetic change. If the lag load rises too high, however, the population becomes extinct. There will be some upper critical value of the lag load, say L_{crit}, consistent with survival. Our question concerning the rate of evolution then becomes: for a given value L_{crit}, how much more rapidly can a sexual population change than an asexual one?

It is not easy to answer this question. A very approximate answer is as follows (Maynard Smith, 1974). If R is the ratio of the rate at which new mutants are incorporated in sexual as compared to asexual populations, then in a small population, $R \simeq 1$. For large populations ($N \geqslant 1/\mu$, where μ is the rate of favourable mutation at a locus),

$$R = \frac{1}{4s} \log (W_{max}/\bar{W}),$$

where s is the selective advantage per locus.

It is difficult to interpret this, because we have little idea of the value of $W_{max}/\bar{W}$. However, it is clear that sexual populations can, when selection is intense, evolve very much more rapidly than asexual ones. This difference will be greatest when a large number of different favourable genetic substitutions are possible, each by itself relatively unimportant.

D   Genetic load, extinction and the Red Queen

In the last section, the basic assumptions were that if conditions change too rapidly the genetic load would increase, and that if the load became too great the population would go extinct. There are many population biologists who have come to regard the concept of a load as misleading, and who in consequence are likely to reject the whole approach. Are they correct?

In Chapter 4, section D, I argue that the taxonomic distribution of parthenogenetic forms leads inescapably to the conclusion that populations committed to parthenogenesis are more likely to go extinct; in saying this, I am only following the view propounded by Stebbins (1950), White (1973 and earlier), Darlington (1939), and others. If the conclusion is accepted, it follows that some populations go extinct because their genotype is not sufficiently well adapted to the contemporary environment, although, given a different genetic system, they could have adapted. It is surely reasonable to say that the difference between the actual genotype and the better-adapted genotype which could have evolved represents a genetic 'load'; it is this load which I have called the lag load. Why then has there been such hostility to the concept of a load?

The idea of a substitutional load traces back to Haldane's (1957) concept of a 'cost of selection'. Here the analogy is with animal breeding. A breeder who wishes to make his breed homogeneous for an initially rare mutant has no choice but to cull many individuals who do not carry the mutant. Haldane's contribution was to show that, if selection was not intense, the total number of individuals lacking a new mutant who must die, or fail to breed, before that mutant is fixed is in the region of ten to thirty times the population size. This, he suggested, placed a limit on the rate of evolutionary change; I have no doubt that he was correct.

As the theory of genetic loads was developed, particularly by Crow & Kimura (1970 and earlier), opposition arose for two reasons, one largely semantic and the other partly emotional. The semantic difficulty has already been touched on; it is surely ridiculous to say that a 'load' has increased when a favourable mutation occurs. I believe that this difficulty disappears if we use the lag load; the load was there already before the mutation occurred, and will be diminished when the mutation is fixed.

The real hostility to the concept arose when it was used to support the view that some allelic variation is selectively neutral. Lewontin & Hubby (1966) argued that the amount of enzyme polymorphism was too great to be maintained by selection favouring heterozygotes unless the selective advantages per locus are very small. Kimura (1968) argued that the apparent rate of molecular evolution was too great to be compatible with Haldane's cost, and hence that most substitutions must be neutral. Many geneticists seem to have felt that, if the load argument could be used to support the neutralist heresy, there must be something wrong with loads.

Two main criticisms of the load argument were made. The first concerns the way in which costs at different loci should be combined. Haldane (1957) had assumed that fitnesses are multiplicative; in this case loads can be added. It was pointed out by a number of people (King, 1967; Milkman, 1967; Sved, Reed & Bodmer, 1967, in relation to segregational loads; Sved, 1968 and Maynard Smith, 1968*b*, in relation to substitutional loads) that a given total intensity of selection (i.e. a given load) can maintain a much greater number of polymorphisms, or cause a greater number of gene substitutions, if the multiplicative assumption is dropped. In particular, this is so if one supposes that selection acts simultaneously on all loci, picking out those individuals who, summed over all loci, have the greatest fitness. This can be called 'threshold' selection, since all individuals with a genetic score above some threshold survive.

The second main criticism (Brues, 1964; Wallace, 1968) is concerned with the way in which selection acts, in particular with the distinction between 'hard' and 'soft' selection. If a particular genotype will always be killed by a particular selective agent (e.g. a particular temperature), this is hard selection. It is reasonable to speak of a cost since the action of the agent reduces the population. In contrast, suppose that the environment is capable of supporting only some fixed number of individuals. Then supernumerary individuals will die in any case. If the ones which die are genetically different from the ones that survive, this constitutes selection, but it is soft selection. There is no cost, since the population is not decreased by the selection. In this latter case, fitnesses will be frequency-dependent; the argument is therefore related to the argument that frequency-dependent selection can maintain polymorphisms without a genetic load (e.g. Clarke, 1972).

The two criticisms are of course related. Victory in intraspecific competition is likely to be determined by the overall effect of genes at many loci (i.e. 'threshold' selection); in general, intraspecific competition leads to soft selection. Hence, 'soft' selection will tend to be 'threshold' selection and 'hard' selection to act independently on different loci, although these associations are by no means absolute. In most cases, selection will be somewhere between hard and soft and somewhere between multiplicative and threshold.

The interpretation of genetic loads is certainly modified by these criticisms, but I do not think that the whole concept is invalidated. Two things must be said about threshold selection. First, with threshold selection the cost required to bring about a number of gene substitutions is indeed reduced, but it does not vanish. Secondly, there is no evidence that in natural populations selection does in fact act in this way, although one can think of situations in which it may well do so. In particular, if one supposes that a species is simultaneously adapting to changes in its pathogens, its predators, its competitors and to climatic conditions, it is hard to see how selection could act in a threshold manner.

The distinction between hard and soft selection is highly relevant to the problem of extinction, and its relation to the genetic load. Thus, suppose that the main evolutionary changes occurring in a species affected characteristics (e.g. size, weapons, behavioural strategies) enabling individuals to occupy dominant positions in a hierarchy and so to increase their chances of survival and breeding. Since there would be genetic differences in fitness, there would be a lag load associated with the changes. Yet the changes which occurred would not make the species less likely to go extinct; indeed precisely the opposite might be the case, because weapons evolved for intraspecific contests may be a handicap in other contexts (if they were not, hinds would have antlers). A parthenogenetic population which made these changes more slowly or not at all would be no more likely to go extinct.

Clearly, the whole concept of lag load and its relation to environmental change and extinction is irrelevant when the selective agent is intraspecific competition of this kind. But this does not invalidate the concept when applied to evolution in response to changes in the physical environment, or in diseases, predators, or competing species.

I cannot end a discussion of the relationship between evolutionary rate and extinction without considering the ideas of Van Valen (1973). He presents evidence for a 'law of constant extinction', according to which, for any group of related organisms with a common ecology (e.g. carnivorous mammals or bony fishes), there is a constant probability of extinction of any taxon (e.g. a genus or family; the fossil record is too incomplete to measure species extinction) per unit time. In an attempt to explain this empirical law, he appeals to the hypothesis of the 'Red Queen', according to which each evolutionary advance by any one species is experienced as a deterioration of the environment by one or more other species. Like the Red Queen, each species must therefore evolve as fast as it can merely to survive.

In general terms, the hypothesis is an attractive one. If it is correct, we would expect to find parthenogenetic and selfing species to be rarest in tropical environments and commonest in temperate and sub-arctic ones, since the challenge from other species is likely to be most intense in the former. Evidence that this is so is discussed in Chapter 6, section D.

There is, however, a difficulty in deriving a law of constant extinction from the Red Queen hypothesis. Van Valen suggests that the derivation depends on a 'zero sum' condition, according to which the increase in fitness of one species is exactly equal to the sum of decrements of fitness of all other species. In fact it can be shown (Maynard Smith, 1976a) that a law of constant extinction per unit time follows only if this zero sum assumption is obeyed exactly. It is hard to see why it should be. It seems far more likely that at some times and in some systems the summed decrements of fitness will be less than the increases, and at other times and in other systems the decrements will be greater than the increases.

It follows that an ecosystem can be in two states: in one, the average lag load per species will be decreasing and in the other it will be increasing. Since the rate of evolution is an increasing function of the lag load, in the former systems evolutionary changes will be slowing down and in the latter they will be accelerating. These systems may be called 'convergent' and 'divergent'. In a divergent system, species extinctions will for a time increase in frequency, and in a convergent one they will decrease.

What then of the law of constant extinction? I think such a law may follow from an argument similar to MacArthur & Wilson's (1967) equilibrium theory of island biogeography. The number of species on an island is an equilibrium between immigration and extinction. The number of species (or genera) in an ecosystem is an equilibrium between extinction and speciation. Since the earth is finite, the total number of individuals at any trophic level will remain approximately constant (except, as Stebbins has pointed out, when a major advance in primary production takes place). If, for whatever reason, Preston's (1962) canonical species abundance distribution is preserved despite changes in species composition, then not only the number of individuals but also the number of species and their abundance will also approximate to the same steady state. Associated with this abundance distribution there will be a rate of chance extinction of the rarer species. It will, of course, not be a matter of chance which species are rare enough to be at risk.

A system of this kind would have some kind of stability. If the number of species rose above the canonical number, extinctions would be more frequent. If it fell below the canonical number, extinctions would be less frequent and, since ecological niches would be vacant, speciation would accelerate. There is of course a correspondence between these two states and convergent and divergent ecosystems; if the number of species is high and species packing close, it is more likely that evolutionary advances by one species will be detrimental to others, and so more likely that the system will be divergent; when species number is low and ecological niches vacant, evolutionary advances by one species will be less detrimental to others, and the system will be convergent.

These ideas are vague and speculative. But it seems clear that there must be some large-scale and long-term stability of ecosystems in evolutionary time.

E   Transient and steady-state models of evolution

The models analysed in this chapter confirm that populations with recombination can evolve more rapidly than ones without, the main exception being small populations in which each favourable mutation is fixed before the next occurs. The effect seems to be large enough to account for the observed pattern of extinction. Yet it is still worth asking which type of model best represents actual evolutionary changes.

For sexual populations, the contrast between transient and steady-state models is illustrated in Figure 6 (Maynard Smith, 1974), which compares two extreme models of evolutionary change in a sexual population. In Figure 6a, it is supposed that selection for any phenotypic character or group of characters is intermittent in magnitude or direction. Three periods are illustrated:

Period 1. Intense directional selection for a group of characters 'X' reduces the genetic variance V of these characters; because of linkage and pleiotropism there will be a small reduction in the variance V_R of characters not under directional selection.

Period 2. Directional selection relaxed. The variance V is restored; the restoration is ultimately caused by new mutation, but there may be changes in the secondary effects of genes so that new genes come to affect character X.

Period 3. Intense directional selection for character Y.

According to this model the response of sexual populations to changes in the environment resembles the response of a captive population to artificial selection in that it depends on pre-existing genetic variance. A model of this kind is appropriate only if periods of

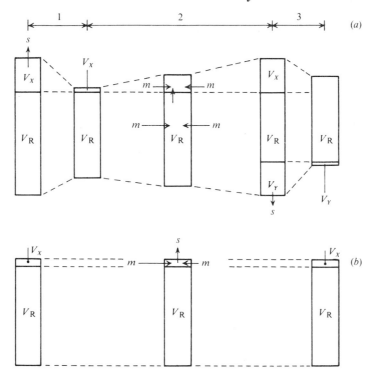

Figure 6. Models of evolution in a sexual population: (*a*), intermittent selection; (*b*), long-continued selection. V_X, V_Y, genetic variance for characters X and Y: V_R, residual genetic variance: *m*, mutation: *s*, selective fixation. (From Maynard Smith, 1974.)

directional selection are separated by much longer periods of relaxed or normalising selection.

An alternative model is shown in Figure 6*b*. It is supposed that there is long-continued directional selection for a single group of characters. The additive genetic variance of the selected characters will be small, and there will be a short-term balance between new mutation and fixation by selection. I shall refer to this as a steady-state model.

Which of these is the more appropriate model? It is hard to say more than that the truth must lie somewhere in between. Another way of looking at the distinction is as follows. If Figure 6*b* is correct, then gene loci could be divided into two categories. There are loci which are polymorphic within species, and which contribute to V_R. Polymorphisms at these loci tend to be very long lasting. There is a second class of loci at which occasional relatively rapid gene substitutions take place; polymorphisms at these loci are transient. There may of course be a third category of loci which do not vary (e.g. the famous histone IV locus, which differs by only one amino acid between higher plants and animals).

The alternative picture, if Figure 6a is correct, is that the normal pattern of gene substitution is for a new mutation to be first established in the population as a stable polymorphism, to remain at an intermediate frequency for a long period, and finally to go to fixation when environmental conditions change.

One reason for wishing to know which of these pictures is nearer the truth is as follows. If fixations, when they occur, are rapid, then the hitch-hiking phenomenon (Maynard Smith & Haigh, 1974; Thomson, 1977; Strobeck, Maynard Smith & Charlesworth, 1976) is important; if Figure 6a is correct, most new alleles have time to reach linkage equilibrium before they are driven to fixation, and hitch-hiking is unimportant.

A second question whose answer hinges on our choice of model concerns the role of the mutation rate in determining the rate of evolution. Thus, if Figure 6a is correct, a doubling of the mutation rate would not greatly alter the rate of adaptive evolution. If Figure 6b is correct, a doubling of the mutation rate would double the rate of adaptive evolution. (Of course, in either case a doubling of the mutation rate would double the rate of neutral substitutions, if such there be.) It is rather disturbing that we should not know the effect on the evolution rate of doubling one of the major parameters.

Figure 6 refers to sexual populations. The corresponding picture for asexual ones is Figure 7. Such a population is composed of a number of clones. Figure 7a shows the response of an asexual population to intense directional selection for character X. Such selection is equivalent to selecting only that clone with the optimal genotype; it eliminates not only the genetic variance for X, but all genetic variance. It will be clear that evolution in an asexual population normally consists of changes within individual clones. If the environment changes, a clone can only adapt to that change by

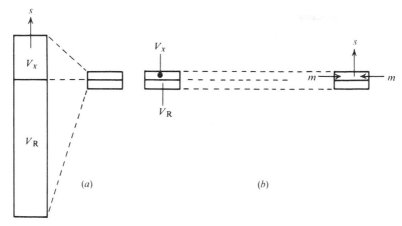

Figure 7. Models of evolution in an asexual population. (*a*) Intermittent selection; (*b*) long-continued selection; notation as in Figure 6. (From Maynard Smith, 1974.)

incorporating newly occurring mutations. The appropriate model is Figure 7*b*.

How is this conclusion to be reconciled with the fact that many parthenogenetic complexes consist of a large number of genetically different clones? It seems likely that these clones are in most cases derived separately from sexual ancestors. Their coexistence is possible only because they are adapted to slightly different ecological niches. Suppose, however, that there were strong directional selection for the same genetic substitution in all clones, for example because of the appearance of a new pathogen. Response to this selection would eliminate most of the genetic variability.

The implication is either that intense selection for a specific substitution is a rare event, or that the generation of new clones from sexual ancestors is frequent. It would, therefore, be intesting to look electrophoretically at parthenogenetic varieties such as *Houttuynia* (p. 65) which have no sexual relatives.

F   A review of the models

Although a number of very different models have been considered in this chapter, it is possible to generalise about the way in which recombination influences evolution. Felsenstein (1974) makes the interesting point that those authors who, explicitly or implicitly, allow for chance events in finite populations conclude that recombination accelerates evolution, whereas those who consider infinite population models (such as that discussed on pp. 13–15) conclude that it does not. This, he suggests, is because of what he calls the 'Hill–Robertson effect' (from a paper by Hill & Robertson, 1966). Selection on alleles at one locus produces gene frequency changes at other loci; the smaller the population, and the lower the rate of recombination between the selected locus and other loci the greater are the effects. In the complete absence of recombination, an increase in frequency of an allele at one locus completely prevents increases in frequency at another, as in the original models of Fisher (1930) and Muller (1932). These effects arise in finite populations, because chance linkage disequilibria are inevitable.

3 Some consequences of sex and recombination – II. Muller's ratchet

The last chapter discussed the effect of recombination in accelerating the accumulation of favourable alleles in a population. In this, I consider the possiblity that recombination may delay the accumulation of harmful alleles. The idea was first suggested by Muller (1964), who pointed out that 'an asexual population incorporates a kind of ratchet mechanism, such as it can never get to contain, in any of its lines, a load of mutation smaller than that already existing in its at present least-loaded lines'. The point is obvious enough, once stated, but was overlooked by population geneticists, until Felsenstein (1974) resurrected it and worked out some of its consequences.

In what circumstances will Muller's ratchet operate?

Consider a haploid asexual population of size N, with generations separate. Each individual has l loci, at each of which there is a probability u per generation that a deleterious mutation will occur; l is large and u small. An individual carrying k deleterious mutations has a fitness $(1-s)^k$. We are interested only in mildly deleterious mutations, so s is in general small. Fitnesses are assumed to be multiplicative.

In any generation, we can classify the individuals according to the number of harmful mutations they carry. Let n_i be the number of individuals with i harmful mutations. Then

$$\sum_{i=0}^{l} n_i = N. \tag{3.1}$$

Muller's ratchet then works as follows. Suppose that the smallest number of deleterious mutations in any individual is j; then $n_0 = n_1 = \ldots n_{j-1} = 0$. In the absence of recombination, there is no way in which an individual with less than j mutations can ever arise (it is shown below that back mutation is effectively irrelevant). However, if n_j is small, there is always a chance that all n_j individuals may die or fail to reproduce. If this happens, then there is a new optimal class, with $j+1$ mutations; the ratchet has clicked round one notch.

Whether or not the ratchet operates depends on the size of the optimal class; if it is small, deleterious mutations accumulate; if it is large they do not.

The following mathematical treatment has been suggested by Dr John Haigh. We first consider a population so large that changes in the n_i can be treated deterministically. We ask what is the 'deterministic equilibrium' – i.e. the distribution of the n_i which exactly reproduces itself under selection and mutation? It can be shown that this distribution is given by

$$n_k = Ne^{-\theta} \theta^k/k!, \tag{3.2}$$

where $\theta = ul/s$.

Writing $ul = U$, the mutation rate per genome, the size of the optimal class is

$$n_0 = Ne^{-U/s}. \tag{3.3}$$

The mean fitness of the population, relative to a fitness of 1 for a member of the optimal class, is

$$\bar{W} \simeq e^{-U}. \tag{3.4}$$

Notice that the probability that an individual undergoes no new mutations is $(1-u)^l = e^{-U} = \bar{W}$. Thus, the mean fitness depends on the mutation rate but not on the selective disadvantage per mutation; in other words, the mutational load, $(W_{max} - \bar{W})/W_{max}$, is equal to $1 - e^{-U}$, the probability that an individual will suffer at least one harmful mutation.

At the deterministic equilibrium mutations are not accumulating; the ratchet is not operating. For this to be an adequate description of an actual population all that is required is that the size of the zero class, n_0, should be large enough to be treated deterministically. Thus if $n_0 \geqslant 1000$, there is effectively no chance that it will, in any generation, drift to zero; if $n_0 \simeq 100$, the ratchet will move, although slowly; if $n_0 \leqslant 10$, the ratchet will move quite rapidly.

Given n, $U = ul$, and s, it is not difficult to simulate the behaviour of a population in which the ratchet is operating. It turns out that at any instant the distribution of n_i is very similar to the deterministic one. With time, however, the distribution moves steadily towards a higher and higher mean number of deleterious mutations, the optimal class occasionally being lost and replaced by a new optimal class with one more deleterious mutation.

It is not possible to give a general expression for the rate of movement of the population mean; that is, for the rate of genetic deterioration. However, it turns out, as expected, that the crucial factor determining the rate is the magnitude of n_0, the optimal class in the deterministic equilibrium distribution. From equation (3.3), it follows that, to an order of magnitude,

if $U/s \leqslant 1$, the ratchet does not operate;

$U/s \simeq 10$, the ratchet operates in any except very large populations; and

$U/s \geqslant 20$, the ratchet operates.

The possible significance of these values will be discussed in a moment. First, however, I want to consider back mutation and then to ask what would be the situation in a population with recombination.

It is easy to see that back mutation is irrelevant. If the ratchet is operating, the number of individuals in the optimal class in any generation varies from one to several hundreds. The number of generations before the class drifts to zero is seldom more than one hundred. Thus the number of individuals in the class, summed over all generations from the time it became the optimal class to the time it is lost, is rarely more than 10^4. Now, back mutation is relevant only if a descendant of one of those 10^4 individuals carried a back mutation, and thus reconstituted a new optimal class with one fewer mutation. Since back mutations occur at a rate of 10^{-8} or less, the process can safely be ignored.

What is the position if there is recombination? At each of l loci there will be a frequency $p = u/s$ of deleterious mutations. With free recombination, these will occur independently, or nearly so, so that the frequency of individuals with $0, 1, 2 \ldots$ deleterious mutations is

$$e^{-ul/s}(1 + (ul/s) + \tfrac{1}{2}(ul/s)^2 + \ldots).$$

That is, it is exactly the same as the deterministic equilibrium distribution for the asexual population. Now, however, there is no ratchet operating. Thus two individuals with k mutations each will, in general, carry different mutations, and so, by recombination, can produce offspring with fewer (and with more) than k mutations.

There is, however, an upper limit on the mutation rate. If, for example, we consider organisms reproducing by binary fission, and assume that selection is acting on survival rather than division time, then $\bar{W} > \frac{1}{2}$; since, from equation (3.4), $\bar{W} \simeq e^{-U}$, it follows that $U < 1$. For organisms capable of producing many offspring the mutational load could, in principle, be greater than this, although it is doubtful if it often is so.

Thus, for a sexual population, the condition which must be satisfied if mutations are not to accumulate is, very roughly, $U < 1$. For an asexual population it is, even more approximately, $U/s < 1$. It follows that the advantages of recombination are important only if the majority of deleterious mutations are only slightly deleterious.

Considered at the population level the implication of Muller's ratchet is as follows. If most deleterious mutations are only slightly deleterious, then for a given error rate per site per replication, the maximum amount of essential genetic material in the genome which can be reproduced without steady deterioration is substantially greater for a population with genetic recombination.

There is one pleasing aspect of this conclusion. It is now widely accepted that the genes responsible for recombination evolved in the first place because of their role in DNA repair. What follows from Muller's argument is that recombination itself functions as a form of repair.

4 Could sex be maintained by group selection? The comparative data

Preamble

This chapter is concerned with the following question: is it possible in eukaryotes, and particularly in higher animals and plants, that sexual reproduction is maintained by group selection, in the face of short-term individual selection in favour of parthenogenesis? Before this question can be answered in the affirmative, the following points must be established.

(i) That there is short-term selection in favour of parthenogenesis. If there were no genetic differences between sexual and partheno-genetic females other than that causing their different methods of reproduction, this would certainly be true for most types of life history. The argument was given on pp. 2–3. Possible reasons why in some cases the argument might not hold are discussed in section A of this chapter. Two further things must be shown. First, that partheno-genesis is not accompanied by genetic changes lowering fitness: this question is discussed in section C of this chapter. Secondly, that sexual reproduction does not have genetically favourable conse-quences in the short run sufficient to counterbalance the twofold advantage of not producing males. This, of course, is the main alternative to the group selection hypothesis; it is discussed in Chapters 6 and 7.

(ii) That there is long-term selection in favour of groups which reproduce sexually. The theoretical reasons for expecting this to be so were discussed in the last two chapters. In section D of this chapter I discuss whether the taxonomic distribution of parthenogenesis supports the group selection argument.

(ii) That the group selection argument is quantitatively plausible. As explained on pp. 4–5, the maintenance of a characteristic by group selection requires that each individually favourable mutant which occurs, and which survives the initial risk of chance extinction, must be balanced by a group extinction. Hence, if the group selection argument is to be quantitatively plausible, the origin of new parthenogenetic strains must be a rare event. The problem of how such strains arise is discussed in sections B and C of this chapter. The general upshot of this discussion is that the origin of such strains may well be rare, especially in animals.

Quantitative plausibility depends not only on the frequency of origin of parthenogenetic strains, but also on the frequency of extinction of groups. What, in this context, is a 'group'? This question is taken up in section F of this chapter.

A final and crucial question concerning quantitative plausibility is discussed in section E. If there is stable coexistence, within a population, of sexual and asexual modes of reproduction, does not this prove that there must be short-term selection favouring the sexual mode? This argument from 'balance' (Williams, 1975) is a powerful one, but, for reasons given in section E, it is not an easy one to evaluate.

A   Do parthenogenetic varieties enjoy a twofold advantage?

The basic argument is presented on pp. 2–3. In effect, a parthenogenetic female enjoys a twofold selective advantage because she wastes no energy on producing sons. The argument holds, of course, only if other things are equal; that is, so long as there are not genetic differences affecting fitness between sexual and asexual females. A large part of this book is concerned with whether such genetic differences exist. In this section, I want to discuss the simpler question of whether there are other factors which might invalidate the argument. There seem to be three: isogamy, paternal care, and hermaphroditism.

Isogamy

In many protozoans and green algae there is no size differentiation of the gametes; instead, two morphologically identical gametes fuse. In such cases, there is no immediate twofold advantage: free exchange is no robbery. Thus consider two equal-sized diploid cells, each of which undergoes meiosis to produce four haploid gametes. The gametes then fuse in pairs to form diploid zygotes, each half the size of the original diploid cells. Now consider a dominant mutant gene, A (for apomixis) instead of $+$, which suppresses meiosis and permits the original diploid cell to undergo mitosis, giving two half-size cells, each with a copy of A. But in the absence of A, there would be two half-size cells each with a copy of $+$.

The only advantage that A has over $+$ is that the time taken for the second meiotic division and subsequent fusion is avoided. This would give A a selective advantage over $+$ in a growing and dividing population. But if the difference were expressed (i.e. if allele $+$ caused meiosis) only in conditions of nutrient shortage when normal growth was in any case impossible, allele A would confer no selective advantage. Hence, one would expect to find, in isogamous species, that gamete production occurs at times of nutrient shortage. If so, apomixis would confer no selective advantage.

The main importance of this is that the essential characteristics of eukaryotic sex, meiosis and syngamy, almost certainly were first evolved in an isogamous population. The evolution from isogamy to anisogamy is discussed in Chapter 9, section C. It follows that, when considering the origins of sexual reproduction in eukaryotes, one need not allow for a twofold disadvantage.

Paternal care

It is an assumption of the argument on p. 3 that a parthenogenetic female can produce as many offspring as a sexual one. This is clearly not true for a sexual species with monogamous pairing in which the male contributes to the care of the young. For such species, there would be no necessary twofold advantage for parthenogenesis.

Hermaphroditism

In general, the twofold advantage of parthenogenesis applies to hermaphrodite species just as it does to bisexual ones, but there are complications which require discussion. Consider, first, a hermaphrodite animal species with external fertilisation (e.g. a sea squirt), or, if you prefer, a self-incompatible hermaphroditic or monoecious plant. Such a species will tend to divide the resources it allocates to reproduction equally to eggs and sperm, or to seed and pollen. The reason for this equal allocation is essentially the same as that which led Fisher (1930) to predict a 1:1 sex ratio in bisexual species; it is discussed further on pp. 130–32. It follows that a parthenogenetic individual, needing to allocate no resources to male functions, could provide twice as many eggs or seeds.

The argument applies also to sequential and to simultaneous hermaphrodites. There are, however, some complications:

Internally fertilising hermaphrodites. Consider an internally fertilising hermaphrodite incapable of self-fertilisation, such as *Helix, Cepaea,* or *Lumbricus.* If each individual mated once only, with reciprocal fertilisation, it would pay the individual to allocate most of its available resources to eggs, and to produce only the relatively small quantity of sperm required to fertilise its partner's eggs. In such a case, parthenogenesis would confer only a small advantage.

The weakness of this argument, of course, is that an individual is not guaranteed a single mating. An individual which can mate many times will not increase the number of offspring it produces as a female, but will increase the number produced as a male. There will, therefore, be selection in favour of allocating more resources to organs of locomotion and courtship which increase mating success; it is certainly one's impression that *Helix* has not skimped resource allocation to male functions. It follows that even in the case of internally fertilising hermaphrodites, there is likely to be a substantial advantage to parthenogenesis, possibly a twofold one.

Self-fertile hermaphrodites. Many hermaphrodite animals and plants are capable of self-fertilisation. If there were no genetic disadvantages associated with self-fertilisation it is clear that such species would rapidly evolve towards this state and many have in fact done so. Once self-fertilisation is the rule, resource allocation to male functions can be reduced; there would then be little further gain from parthenogenesis. In effect, self-fertilising hermaphrodites obtain most of the twofold advantage of parthenogenesis; they also share with parthenogens the ecological advantage that a single individual can colonise a new area. The advantage which at least some types of parthenogenesis have over self-fertile hermaphroditism is the possibility of perpetuating a heterozygous genotype.

The origin and consequences of male sterility. Suppose that a parthenogenetic variety arises within a population of externally fertilising hermaphrodites. Let the gene or genes responsible for the parthenogenetic development of the eggs (or seeds) be called P. There is no reason why the genes P should also cause male sterility. If, in fact, parthenogenetic individuals continue to allocate as much resources to sperm (or pollen) production as do sexual ones, then they do not obtain the twofold advantage.

Nevertheless we would expect genes P to increase in frequency, for the following reason. If parthenogenetic individuals produced the same number of eggs as sexual ones, and produced no pollen, then the frequency of genes P would be constant. Genes P, however, will also be transmitted in pollen to the offspring of sexual individuals. Hence the parthenogenetic variety will increase in frequency at the expense of the sexual ones. Once the parthenogenetic variety has been established as the predominant one in any locality, there will be strong selection for male sterility and the reallocation of resources to female functions.

It is a striking fact that parthenogenetic varieties of plants often continue to allocate substantial resources to male functions. Asexual clones of the dandelion *Taraxacum officinalis* continue to produce functionless yellow petals and many produce functionless pollen. It is difficult to suggest any explanation of these facts, other than that these clones may be relatively recent in origin, and that evolutionary adaptation in asexual populations is slow, so that maladapted features are retained (see p. 54 for a more detailed discussion of *Taraxacum*).

Before a parthenogenetic clone derived from a hermaphrodite ancestor can obtain the twofold advantage, it is necessary both that male sterility should evolve and that the resources previously allocated to male functions be re-allocated to female ones. It seems likely that a genetic change causing male sterility would also cause resource re-allocation, without need for further genetic change. Whether this is so or not would depend on how resource allocation in the individual is controlled. The reason for believing that re-allocation would follow automatically on male sterility comes from the study of gynodioecious plants; as pointed out by Darwin (1877), male sterile plants produce more seed than hermaphrodite ones. The argument is not decisive, because it is possible that gynodioecious species have acquired separate genes for male sterility and resource re-allocation.

In conclusion, the twofold advantage of parthenogenesis does not apply to isogamous species, or to monogamous species with paternal care of the young. In general, it does apply to parthenogens derived from hermaphrodites, but with reservations. It does not apply if the parental population consists of self-fertilising hermaphrodites; the full twofold advantage may only accrue after subsidiary genetic changes have taken place, and this does not always happen.

In the context of the argument concerning the maintenance of sexual reproduction in higher eukaryotes the reservations discussed in this section are relatively unimportant. They do not alter the basic problem of explaining the maintenance of sex in the face of a twofold selective advantage of parthenogenesis.

B   The nature, genetic consequences and origins of parthenogenesis in animals

By parthenogenesis is meant the development of a new individual from an unfertilised egg. It takes two main forms:

(i) 'Arrhenotoky', the production of males from unfertilised eggs, as in the haplo-diploid system found in Hymenoptera and elsewhere, in which the haploid males develop from unfertilised eggs and the diploid females from fertilised ones. A population adopting arrhenotoky does not lose the long-term evolutionary advantages of sex; the main interest of the system in the present context is that it provides a control which can be compared to other forms of parthenogenesis.

(ii) 'Thelytoky', the production of females from unfertilised eggs; may be obligate, cyclical, facultative, or a rare aberration.

It is at the outset essential to distinguish two types of thelytoky, apomictic and automictic, which have opposite genetic consequences. In apomixis, meiosis is suppressed, and there is a single mitotic maturation division, the offspring being genetically identical to their mother. In automixis, meiosis is normal, producing four haploid pronuclei. The diploid number is then restored by the fusion either of two pronuclei or of two early cleavage nuclei. Its exact genetic consequences depend on the way in which diploidy is restored, but in all cases the offspring differ from and are more homozygous than their mother.

It is known that a wide variety of non-specific stimuli will, in many animals, cause an unfertilised egg to start development. The development is often abortive because of the haploid nature of the embryo. It is not uncommon, however, for virgin females of typical bisexual species to lay a few eggs which hatch (e.g. in the stick insects, *Drosophila* species, grouse locusts, grasshoppers; references in White, 1973, p. 701). Carson (1967) was able by selection to increase the frequency of hatching of unfertilised eggs in *D. mercatorum* from 0.01 to 0.06, and in this way to obtain a permanently thelytokous strain of a typically bisexual species. However, in all the cases which have been studied, parthenogenesis of this sporadic kind has turned out to be automictic. In *D. mercatorum* the diploid chromosome number is usually restored by fusion of cleavage nuclei, leading to immediate homozygosity at all loci.

Not surprisingly, natural thelytokous populations of this kind, with enforced homozygosity in one generation, are extremely rare. Nur (1971) makes the interesting point that most of the very few known cases are probably descended from arrhenotokous ancestors. In typical diploid outbreeding sexual species, the great majority of genotypes made homozygous at all loci would be lethal or nearly so. In haplo-diploid species, the load of lethal and deleterious recessives will be much lower, because all genes are exposed to selection in haploids, and consequently there would be a much better chance of an arbitrary homozygote having high viability. Nur states that the only reported cases of natural populations adopting this type of automixis are a few species of coccid bug, one species of white fly and one species of mite; in each case the nearest bisexual relatives have effectively haploid males. A possible exception to Nur's generalisation is the stick insect, *Bacillus rossius,* which is bisexual in North Africa but in France is automictic with endomitotic doubling in the embryo (Pijnacker, 1969).

It follows that, even if sporadic parthenogenetic development of eggs is not all that rare, such parthenogenesis is usually automictic and associated with enforced homozygosity, and consequently an unpromising starting point for the origin of a parthenogenetic variety.

Automixis in which diploidy is restored by fusion of two pronuclei is less rare in natural populations than the type just discussed. It also causes homozygosity, but not immediate homozygosity at all loci. The detailed genetic consequences are shown in Figure 8. The important point is that the pronuclei are often arranged in the egg in a regular way, so that the egg nucleus can be formed always by a pair of sister nuclei, or always by a pair of non-sister nuclei. This fact can be exploited to enable an automictic species to maintain heterozygosity at least at some loci. For example, the only naturally parthenogenetic *Drosophila* species, *D. mangabeirai,* is automictic, with two non-sister nuclei fusing to form the egg nucleus. All individuals are diploid, and heterozygous for three inversions; presumably, few chiasmata form

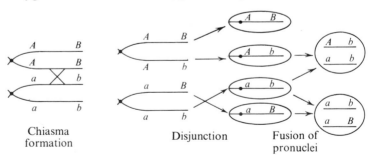

Chiasma formation   Disjunction   Fusion of pronuclei

Figure 8. Genetic consequences of automixis. If two centrally placed, non-sister pronuclei fuse, this preserves heterozygosity for loci between the centromere and the first chiasma *(Aa),* but may cause homozygosity for distal loci *(bb).* If two sister pronuclei fuse, this causes homozygosity for proximal loci.

between the centromere and these inversions. Even with this rather specialised type of restitution, egg mortality in *D. mangabeirai* is high (approximately 40%).

Another species which exploits the genetic advantages of restoring diploidy by the fusion of non-sister nuclei is the moth *Solenobia triquetrella* (Seiler, 1963). In addition to a bisexual diploid race, there are diploid and tetraploid races with this type of thelytoky. The importance of this case is that in the Lepidoptera the female is the heterogametic sex, so that a mechanism leading to homozygosity would give XX (male) and YY (probably lethal) offspring. In birds, where again the female is the heterogametic sex, this difficulty has so far prevented the establishment of permanent thelytokous strains by artificial selection. Olsen (1965) was able greatly to increase the frequency with which unfertilized eggs of a strain of turkeys hatched, but the hatchlings were usually males.

A comparison of the situation in turkeys with that in *S. triquetrella* or of *D. mercatorium* with *D. mangabeirai,* shows that the origins of a functional and effective mechanism of automictic parthenogenesis is not a simple matter, and hence is unlikely to be a common event.

Not all automictic mechanisms, however, guarantee the fusion of non-sister pronuclei. For example, in the enchytraeid oligochaets, thelytoky when it occurs is usually by the suppression of the second maturation division (Christensen, 1961), a process which is genetically equivalent to fusion of sister pronuclei. It would be expected to produce homozygosity for all loci between the centromere and the first chiasma. It is not clear whether enchytraeids can tolerate homozygosity or whether they have some device for retaining effective heterozygosity; it may be relevant that most thelytokous strains are even-numbered polyploids. Studies of enzyme patterns in these and other automictic populations would be interesting.

In contrast to automixis, the characteristic feature of apomixis is that genetic heterozygosity is maintained, and may indeed be increased by mutation after the establishment of parthenogenesis. White (1973 and earlier), Stebbins (1950) and others have argued that this ability to transmit unchanged a well-adapted heterotic genotype is the main reason for the evolutionary success of apomicts. Apomixis is widely distributed in animals. In weevils (the only group of beetles in which thelytoky is at all common) it is the only mechanism which has been described. It is also known to occur, for example, in hymenopterans, coccids, dipterans, orthopterans, crustaceans, nematodes, and oligochaets. This list is not intended to be exhaustive. Indeed, it is important to remember that the great majority of thelytokous species have not been investigated cytologically; oogenesis is not the easiest process in which to examine chromosome behaviour.

One final mechanism of thelytoky must be mentioned, if only because it occurs in the most successful group of parthenogenetic vertebrates. There are some thirty known races of parthenogenetic lizards, from five genera, including wall lizards *(Lacerta)*, geckos *(Hemidactylus)*, and whiptails *(Cnemidophorus)* (Maslin, 1971). In *Cnemidophorus*, Cuellar (1971) has shown that there is a pre-meiotic doubling of the chromosomes (endomitosis), followed by an apparently normal meiosis with the diploid (or in some races, triploid) number of bivalents, giving rise to an egg nucleus with the parental chromosome number.

The genetic consequences of this process depend on the way in which the bivalents are formed. If, as seems likely, only sister chromosomes pair, the process is genetically equivalent to apomixis. Offspring will be genetically identical to their parents, and heterozygosity will be maintained. If, on the other hand, non-sister chromosomes form bivalents, homozygosity will increase. The main reason for suspecting that only (or mainly) sister chromosomes pair is that some races of *Cnemidophorus* consist entirely of individuals heterozygous at enzyme loci and for chromosome rearrangements (Neaves, 1969).

Meiotic parthenogenesis in which the chromosome number is maintained by pre-meiotic doubling rather than post-meiotic fusion is also known in invertebrates. It occurs in the Australian grasshopper *Moraba virgo* (White, 1966), and in parthenogenetic earthworms. It is interesting that *M. virgo*, like some races of *Cnemidophorus*, is heterozygous for chromosome rearrangements.

The nature of parthenogenesis in the other lizard genera is uncertain. Apart from the lizards, the only known cases of naturally occurring parthenogenesis in vertebrates are in the bony fishes

Poecilia and *Poeciliopsis* and in the salamander *Ambystoma* (Cuellar, 1974). In *Ambystoma*, there are two diploid bisexual species, *A. laterale* and *A. jeffersonianum*, and two triploid thelytokous forms, one having two *laterale* and one *jeffersonianum* chromosome sets and the other one *laterale* and two *jeffersonianum* sets; this interpretation has been confirmed by studies of serum proteins (Uzell & Goldblatt, 1967). The mechanism of thelytoky is again pre-meiotic doubling followed by meiosis. An interesting additional complication is that the thelytokous females must usually mate with a male of one of the sexual species before they produce fertilised eggs. Apparently the sperm penetrates the egg, thus initiating cleavage, but contributes no genetic material. It seems, however, that these races may be on the way to emancipating themselves from dependence on the sexual species, since all-female thelytokous populations occasionally occur in localities by themselves.

'Gynogenesis', in which a female produces eggs parthenogeneti-cally only after mating with a male, occurs sporadically among invertebrates. It is also a feature of parthenogenesis in the fish *Poecilia* and *Poecilopsis*. *Poecilia formosa* is a diploid thelytokous species which is apparently a hybrid between two sympatric sexual species. The type of thelytoky is unknown, but development of the egg is entirely dependent on mating with males from one of these sexual species.

An even more bizarre situation exists in *Poeciliopsis* (Schultz, 1969, 1973). Two sexual species, *P. lucida* and *P. monacha* have given rise to two triploid thelytokous forms in a way exactly analogous to that described above for triploid *Amblystoma*. There are, in addition, diploid all-female biotypes intermediate between the two bisexual species. These females mate with males from a sexual species, and produce female offspring in which both their own genes and those of the male are expressed. At meiosis, however, the male genes are eliminated and only the female genes transmitted.

What can we say of the origins of parthenogenesis? As we have seen, automictic parthenogenesis occurs sporadically in many organisms, and its frequency can be increased by artificial selection. The genetic consequences of automixis, however, are such that automictic strains are unlikely to be established in nature until further genetic changes have taken place which prevent homozygosity or mitigate its adverse effects.

It is less easy to suggest how apomictic thelytoky, or meiotic thelytoky with pre-meiotic doubling, could originate. The association between thelytoky and species hybridisation in vertebrates *(Poecilia, Poeciliopsis, Amblystoma,* some races of *Cnemidophorus)* has led to the idea that hybridisation, by bringing together incompatible chromosome sets, which either fail to form proper bivalents or which fail to disjoin, may be a cause of parthenogenesis. This view has been cogently criticised by Cuellar (1974). He points out that the abnormal behaviour of chromosomes in meiosis could hardly be a cause, in the physiological sense, of endomitosis, which occurs *before* meiosis. He favours instead the view that genetic modifiers first accumulate to cause endomitosis and parthenogenetic development, giving rise to an autodiploid race. Hybridisation between such a race and males from a different sexual species could then give rise to a triploid race; the modifiers already present in the diploid would ensure the parthenogenetic reproduction of this race.

The parthenogenetic races of *Amblystoma* and *Poeciliopsis,* and the triploid biotypes of *Cnemidophorus,* could well have arisen in this way. Indeed, as White (1973) points out, the only way one can easily envisage a triploid variety arising is by the fertilisation of a diploid egg by a sperm from a haploid male. Some *Cnemidophorus* varieties probably represent the autodiploid stage of this process; for example, *C. lemniscatus* exists in diploid sexual and thelytokous varieties which are chromosomally and morphologically indistinguishable, in an area where no other bisexual species exist (Vanzolini, 1970). There are, however, two diploids, *C. neomexicanus* and *C. cozemula,* which are chromosomally heteromorphic and are thought to be hybrids between existing bisexual species. The latter varieties are hard to explain on Cuellar's view that thelytoky originates within a species.

There seems to be no doubt that *Poecilia formosa* is a species hybrid. Cuellar points out, however, that when the two presumed parental species are crossed, hybrids are produced which are indeed indistinguishable from *P. formosa* morphologically, but which are of both sexes, and never parthenogenetic. Hybridisation is not by itself sufficient to cause parthenogenetic development.

The facts concerning parthenogenesis in animals are bewilderingly hard for a non-cytologist to digest and understand. In the preceding account, I have relied heavily on White (1973) and on the review papers in the *American Zoologist* (1971), **11.** The impression with which I am left is that, while it is rather a common event for a female of a normally sexual species to lay an egg which develops without fertilisation, it is a very rare event for a variety to arise in which females regularly and reliably produce parthenogenetic offspring of high genetic fitness.

C   Parthenogenesis in plants

An immediate difficulty in comparing parthenogenesis in plants and animals arises because of the wholly different nomenclature used in the two fields. For example, Stebbins (1950) uses the word 'apomixis' to include all forms of vegetative reproduction, whether or not a seed is produced, whereas by 'parthenogenesis' he means a specific form in which a diploid egg cell develops without fertilisation. Thus, for Stebbins, parthenogenesis is a special kind of apomixis, whereas for White apomixis is a special kind of parthenogenesis.

Clearly, there is no way to bring these terminologies into line. Fortunately, it does not matter for our present purpose. At least in flowering plants, all methods of asexual reproduction known to occur in nature have the same genetic consequence: offspring are produced which are genetically identical to the parent plant.

This may occur without the production of seeds, by bulbs, bulbils, layers, runners, and so on. It is usual for these processes to occur in plants which can also reproduce sexually. This type of asexual reproduction has only limited relevance to the present argument. Thus, an individual which abandoned seed production entirely would not only lose the genetic advantages (if any) of sexual reproduction; it would also lose its main means of dispersing its offspring. Thus the existence of many species which produce both sexual seeds and asexual runners is not relevant to the 'balance' argument. It is, however, relevant that it is the dispersive stage which is produced sexually, since this is the stage which will meet an unpredictable environment.

We are here concerned mainly with species which produce seeds asexually. This can happen in a number of ways. A diploid embryo may arise from the diploid sporophyte tissue of the parent, without the intervention of any gametophytic stage. Alternatively, a diploid gametophyte may be produced ameiotically, and this in turn can give rise to the embryo from either a (diploid) egg cell or some other tissue of the gametophyte. But all this is genetically irrelevant; the offspring in all cases are genetically identical to the parents.

It is important to remember that the great majority of flowering plants have hermaphrodite flowers, and that apomictic varieties seem always to have originated from such species. This fact has some important consequences. First, it may explain the absence of automictic parthenogenesis; if it were selectively advantageous for a hermaphrodite to evolve into an automict, the same result could be achieved more easily by becoming a self-fertile hermaphrodite, an evolutionary road which has been followed many times.

A second consequence of hermaphroditism is that an apomictic mutant does not automatically acquire the twofold advantage, since the genetic change responsible for parthenogenesis would not suppress pollen production. Hence, further evolutionary changes would be required before the full advantage of asexuality was achieved.

Thirdly, there is the occurrence of 'pseudogamy' in apomictic plant complexes. For example, in the grass *Poa* and most varieties of *Potentilla,* apomictic plants produce functional pollen, and require fertilisation before they set seed. The offspring, however, are maternal in character; the pollen serves only to fertilise the endo-sperm (a tissue of nutritive function in the seed). The phenomenon is reminiscent of the gynogenesis found in *Poecilia* and other animals, but its evolutionary significance is different. In animals, the gyno-genetic varieties are restricted to the range of the ancestral sexual species, and their evolutionary future, if they have one, is to become autonomously parthenogenetic. In contrast, pseudogamy in plants seems to confer greater evolutionary potential than ordinary apomixis; the matter is discussed further on pp. 65–6.

In plants, apomixis is commonly associated with polyploidy, and with evidence of hybrid origin, between species or at least between ecotypes (Stebbins, 1971). There is no reason to think that hybridisation by itself causes apomixis in plants any more than it does in animals. The association with hybridisation probably exists because a hybrid genotype may adapt a plant to a new habitat, so that if apomixis happens to arise in such a plant it has a good chance of surviving. The association with polyploidy may exist because if a polyploid genotype does arise in an apomict it will not cause infertility as it would in a sexual species.

The crucial questions in the present context are:

  (i) How common are new apomictic mutants likely to be?
 (ii) How extensive is facultative apomixis?

The second question is discussed in section E. The former is hard to answer. Asexual reproduction of plants by seed is not faced with the same difficulties as parthenogenesis in animals; in particular, it does not lead to inviable homozygous genotypes. A number of different types of effectively ameiotic apomixis have evolved. It may be relevant here that in at least some higher plants a single somatic cell can give rise to a new plant (Street, 1974), whereas this seems not to be the case in higher animals. The evidence suggests that meiosis is usually a prerequisite for development and differentiation in animals, so that ameiotic parthenogenesis cannot easily arise, whereas in plants there is no causal connection between meiosis and the restoration of totipotence. It is, therefore, likely that new apomictic varieties are commoner in plants than animals, although I know of no direct evidence. If it is true, then the case for a 'group selection' explanation of the maintenance of sex is weaker for plants.

D   The evolutionary potential of parthenogenetic strains

How great an evolutionary change can go on within a parthenogenetic strain? There are two observational approaches to this question: one can look at the taxonomic distribution of existing parthenogens, and one can look at the range of variation within existing parthenogenetic varieties.

It is part of the accepted wisdom of evolution theorists that the sporadic taxonomic distribution of parthenogenetic forms demonstrates that they are evolutionary dead ends, doomed to extinction. I would not think it necessary to rediscuss this conclusion had it not recently been challenged by Williams (1975). Williams makes two points; first, extinction does not always or even usually result from a lack of evolutionary potential, and, secondly, the observed taxonomic distribution could be the result of chance.

On the first point, I cannot make Williams's point better than by repeating his own vivid, if hypothetical, example. Suppose, he says, there had been a species of flea confined to the passenger pigeon. That species would now be extinct. It would not have been saved from extinction by a genetic mechanism better able to ensure rapid evolution, because any evolutionary progress the species did make would only be to adapt it to a way of life doomed to become impossible. In other words, most species go extinct, not because they lack evolutionary potential, but because their niche ceases to exist. They run out of niche.

Clearly, some species go extinct for this reason, and it may be that most do. Nevertheless it is logically possible *both* that most species which go extinct do so because they run out of niche *and* that all parthenogenetic species must go extinct because of lack of evolutionary potential, even if their niche persists. I will return to this question after discussing the problem of taxonomic distribution.

Williams shows that if you simulate a random process of extinction and speciation, you finish up with a phylogenetic tree in which some single lineages give rise to many descendant lineages, so that one would be tempted, if one did not know that survival and extinction had been random, to suggest that some important evolutionary advance had been made by the lineage in question, whereas other lineages failed to make such an advance. How then is one to decide whether the observed distribution of parthenogenetic varieties is consistent with chance extinction? The obvious way seems to be to compare the distribution of thelytokous species with that of arrhenotokous ones; if the genetic system does not influence chances of species survival, the distribution patterns should be the same. They are not.

Existing arrhenotokous species may well be descended from as few as eight ancestral arrhenotokous lineages, of which six were insects. The order Hymenoptera is predominantly arrhenotokous, with a few thelytokous species. The order Thysanoptera (thrips) contains many arrhenotokous species, and is probably similar to the Hymenoptera, although fewer species have been studied. Among homopteran bugs, the tribe Iceryini are arrhenotokous, as are a few white flies (family Aleurodidae). Among the Coleoptera, a few bark beetles of the genus *Xyleborus*, and the species *Micromalthus debilis*, are arrhenotokous; the latter is the only species in its family. Outside the insects, the rotifer order Monogononta is predominantly arrhenotokous, with a few thelytokous species. Finally, the mites (Acarini) contain a number of arrhenotokous families and sub-families; it is possible that the system arose only once in the order. Thus of eight examples, five consist of families or larger taxonomic groups, and a sixth of a taxonomically isolated species.

In contrast, even if one treats those parthenogenetic complexes which consists of a large number of closely related clones as being descended from a single ancestor (and this would probably be wrong – see pp. 54–7), existing parthenogenetic populations must be descended from many hundreds of different ancestral lineages. Yet, with one important exception, in no case does a major taxonomic group (sub-family or above) consist predominantly of thelytokous populations, and there is no taxonomically isolated thelytokous species comparable to the arrhenotokous *Micromalthus*. Instead, most thelytokous populations are very similar to existing bisexual ones. Examples of this have been mentioned above in other contexts. To make the same point in a less anecdotal way, there are twenty-eight known obligate thelytokous varieties of psocids: they belong to thirteen different families, and in twelve of the twenty-eight cases there are sexual and parthenogenetic forms of the same nominal species (Mockford, 1971). A fundamentally similar picture is found in plants.

A striking exception to the general rule is afforded by the bdelloid rotifers. No males have ever been described from this rotifer order. This pattern is so different from that observed in the rest of the animal and plant kingdoms that one cannot help wondering whether bdelloids may not have adopted some other means of exchanging genes. At present no evidence exists to support such a speculation, but further study is certainly called for.

Even when full weight has been given to the bdelloids, there can be no doubt that the taxonomic distribution of thelytokous and arrhenotokous species is wholly different. The facts fully support the conventional view that parthenogenetic varieties are doomed to early extinction. The evidence seems to me so strong that it can be taken as showing that extinction is an almost inevitable consequence of a lack of evolutionary potential. This lends support to the Red Queen theory of Van Valen, according to which each species must evolve continuously if it is to avoid extinction caused by the evolution of its competitors, predators, and parasites.

This Red Queen approach may help to explain an apparent paradox discussed by White (1973). Thelytokous species may have a limited range in time, but often they have a wider geographical distribution than their bisexual competitors (e.g. Mockford, 1971, for psocids; Suomalainen, 1950, for weevils; Stebbins, 1950, for plant examples). A single genotype may have a sufficiently wide range of physiological tolerance to compete successfully over a wide geographical range, helped as it is by the twofold advantage of not producing males, and the further advantage that a single individual can colonise a local habitat. But that same single genotype is unlikely to be resistant to parasites which evolve new biochemical specificities.

A second approach to the question of the extent of evolutionary adaptation within a clone is to look at the range of variation within such clones. Unfortunately, this approach runs into the immediate difficulty that one does not know whether the different biotypes have evolved since the origin of parthenogenesis, or whether they merely reflect a polyphyletic origin.

In plants, apomictic varieties tend to exist not as uniform and isolated populations but as groups of taxonomically related but varying biotypes. The genus *Taraxacum* (dandelions) contains some 2000 apomictic and 50 sexual species. The sexual species are all diploid, mainly self-incompatible, and have relatively limited disjunct distributions in West and Central Asia and the Mediterranean. The apomicts are distributed throughout Europe and Asia and into Arctic North America; one Arctic form is also found in the southern tip of South America and in New Zealand, regions it probably reached in the Cretaceous before the southern continents had drifted apart.

The apomicts are all polyploid, mostly triploid or tetraploid. Some lack pollen, but many species produce reduced pollen. The great majority produce only unreduced eggs, genetically identical to the parent plant and destined to develop without fertilisation, although a few triploids are facultative apomicts, producing both unreduced eggs and reduced eggs which develop only after fertilisation. A number of hybrids have been formed experimentally; hybrids between two sexual species are fertile and sexual, whereas those between a sexual species as female parent and an apomict as pollen parent are fertile apomicts. Hybrids are rare in nature because sexual species rarely overlap with apomicts producing fertile pollen.

Richards (1973), from whose review these facts are taken, believes that *Taraxacum* originated in the Cretaceous, perhaps in the western Himalayas, and soon gave rise to apomicts with a wide geographical distribution. He suggests, however, that the majority of existing apomicts (including the native British dandelion, *T. officinalis*, which has recently been introduced into the United States) have arisen by hybridisation since the end of the last ice age, as sexual species moved northwards into regions occupied by Arctic-adapted apomicts able to act as pollen parents.

This illustrates the complexity of agamic complexes in plants. It is hard to say how much evolution, if any, has occurred within asexual populations, although the southern forms which have long been isolated from the rest of the complex would repay study from this point of view. An obvious puzzle is the continued production of bright yellow flowers, and of fertile pollen, by so many apomicts.

It is characteristic of apomictic plants that they should form species complexes, including apomictic and sexual varieties, with the apomicts usually polyploid and with a wider geographic distribution than the sexuals. Other examples are the *Crepis* species of North America, and *Hieracium* and *Antennaria*. Other apomictic complexes, such as *Rubus* subgenus *Eubatus* (blackberries), *Poa,* and *Potentilla* differ from *Taraxacum* in that many of the apomictic varieties are facultatively apomictic, and capable of exchanging genes; these are discussed further on pp. 65–6. The study of apomictic complexes in plants suggests that little evolutionary change occurs within a clone. Stebbins (1971) states that in all cases the range of variation of apomictic forms is bounded by a group of sexual or facultatively apomictic species which are capable of hybridising and exchanging genes with each other.

There is less general agreement on the origin of variations in apomictic complexes in animals. The simuliid fly *Cnephia mutata* exists as a sexual diploid, and in an enormous variety of triploid parthenogenetic biotypes differing in chromosome structure (Basrur & Rothfels, 1959). The authors prefer the hypothesis that this variation has arisen within the parthenogenetic population, but cannot rule out the possibility of polyphyletic origin.

Although *Cnephia* is an extreme example, at least for animals, the existence of a number of biotypes within a parthenogenetic complex is common. In a few cases it seems certain that the variation has arisen within a clone. For example, *Moraba virgo* (an automict with pre-meiotic doubling – see p. 46) has a typical structurally heterozygous karyotype, but in two small localities additional chromosomal rearrangements have been imposed on the basic pattern (White, 1966). Some evolution therefore has occurred within the clone; as against this, White points out that *M. virgo* is relatively uniform in colour, whereas most morabine grasshoppers are highly polymorphic.

One other argument put forward by White (1973), tending to show that relatively little chromosomal change has occurred within clones, is worth mentioning. Thelytokous dipterans are often heterozygous for inversions. These, however, are always of the paracentric type which are common in bisexual dipteran populations. Pericentric inversions are absent from bisexual populations, because they would lower the fertility of the females carrying them. They would, in general, have no such adverse effect in thelytokous females; their absence therefore indicates that little chromosomal evolution has gone on within thelytokous dipteran populations.

In some cases, there is more direct evidence for close genetic similarity between the members of a parthenogenetic variety. In the lizard *Cnemidophorus uniparens,* Cuellar (1976) has performed skin grafts among twenty individuals of two separate populations. Of these, 98.8% were permanently accepted, a result consistent with all the individuals being genetically identical at the relevant loci.

It may be that *C. uniparens* is of very recent origin. Parker & Selander (1976) have found, using electrophoretic techniques, that another parthenogenetic 'species', *C. tesselatus*, is more variable. This species consists of diploid hybrids between two sexual species, and of triploid hybrids between diploid *C. tesselatus* and a third sexual species. The authors studied enzymes at twenty-one loci. All the triploids examined belonged to a single clone, suggesting a single origin, but twelve distinct diploid biotypes were found. These are most easily interpreted as involving five distinct hybridisation events. Four further biotypes have probably arisen by recombination within a clone. As explained earlier, there is pre-meiotic doubling of the chromosomes; pairing between non-sister chromosomes could then lead to the production of genetic novelties by recombination. Finally, three biotypes appear to have arisen by mutation within a clone; each of these is confined to a single locality. Unfortunately, the age of *C. tesselatus* is unknown; it could be over 10 000 years or as little as a few hundred years.

Lokki *et al.* (1975) have shown that there is a large number of electrophoretically different clones in the moth *Solenobia triquetrella*, but the authors believe these represent different origins rather than within-clone evolution. On the other hand, Suomalainen & Saura (1973) believe that the large number of electrophoretically distinct clones of the parthenogenetic weevil *Otiorrhynchus scaber* have arisen by evolution within a parthenogenetic strain. Their data are consistent with this hypothesis, but do not seem to rule out multiple origins. In both *Solenobia* and *Otiorrhynchus*, parthenogenetic strains have a wide geographic range, whereas the diploid bisexual forms are relict in the Alps.

E   Cyclical and facultative parthenogenesis – the balance argument

The strongest argument in favour of a short-term advantage for sex put forward by Williams (1975) is the 'balance' argument. If parthenogenetic and sexual methods of reproduction coexist in the same population, then there must be some immediate advantage to sexual reproduction to counterbalance its twofold disadvantage. Indeed, one would expect natural selection to alter the relative frequencies of sexual and asexual reproduction until the selective advantages balance out. To put the same argument in a different way, group selection can maintain sex by eliminating parthenogenetic varieties if the latter arise sufficiently rarely, but cannot maintain a balance between the two processes unless group extinction is exceedingly common.

The argument seems decisive, but it turns out to be hard to evaluate in practice. There are three situations relevant to the argument:

(i) *Cyclical parthenogenesis,* in which there is a regular alternation between sexual and asexual reproduction. Here the argument is that if the members of a population are capable of reproducing asexually, mutations causing them to do so always must be reasonably common, and would inevitably be established by individual selection (because of the twofold cost of producing males) unless there were some short-term counterbalancing selection.

(ii) *Facultative parthenogenesis,* in which all (or some) females in a population are able to lay both fertilised and unfertilised eggs which hatch. Here the argument is that mutations altering the relative frequency of the two processes in individual females would occur and would be established by individual selection.

(iii) *Coexistence of sexual and parthenogenetic strains.* Here the argument is that if the sexual strain did not have some short-term advantage it would be eliminated.

The first two of these phenomena are discussed in this section, and the third in section F, below.

Cyclical parthenogenesis

It will be convenient to discuss this first for the water flea, *Daphnia,* a small crustacean abundant in lakes and ponds. All-female populations can reproduce indefinitely. There is genetic evidence, discussed shortly, that parthenogenesis is apomictic. The eggs produced hatch immediately into females. Occasionally, under environmental influences whose nature is uncertain, sexual males and females are produced parthenogenetically. These mate, and the females lay special 'winter eggs' which do not hatch immediately, but fall to the bottom of the pond. Should the pond dry up, all life stages except the winter eggs die. Should the pond then refill with water, the winter eggs hatch to provide a new population.

Young (1976), building on earlier work by Hebert (1974), found that a newly refilled pond contained a population which, for three enzyme loci, was approximately in the Hardy–Weinberg ratio. This is consistent with the view that the population was recently derived from sexually produced winter eggs. After some months, all homozygotes at these loci had disappeared, presumably because of adverse selection, but a large number of different multiply-heterozygous clones were still present. An unexplained feature of Young's results is that these clones continued to coexist for two further years with only slight fluctuations in relative frequency, a fact which it is hard to reconcile with concepts of competitive exclusion.

The first point to emphasise is that the parthenogenetic and sexual eggs have different ecological roles. A strain which produced only parthenogenetic eggs would not survive the temporary drying up of its pond, and would have lost the main dispersal stage able to colonise new ponds. This, at first sight, weakens the balance argument. Before sex can be abandoned, what is required is not a mutation which suppresses the sexual stage, but one which enables females to produce winter eggs parthenogenetically.

Such genetic changes are not uncommon. Banta (1925) reported an all-female Long Island population of *D. pulex* which was able to produce winter eggs parthenogenetically. More recently, Zaffagnini & Sabelli (1972) described the apomictic development of winter eggs in the Arctic species *D. middendorffiana,* in which it is typical. These authors review other findings of this kind, and conclude that the production of winter eggs in the absence of males is common in Arctic species, and occurs sometimes in more southern populations.

One is, therefore, faced with two alternative explanations of cyclical parthenogenesis in *Daphnia*. The 'group selection' explanation is that strains such as *D. middendorffiana* will go extinct. When such strains arise, the advantage they obtain by not producing males enables them to replace their sexual competitors. They originate very seldom. The 'balance' explanation is that such wholly parthenogenetic mutations occur frequently, but usually fail to establish themselves because of some short-term disadvantage; strains such as *D. middendorffiana* are unusual in that the wholly parthenogenetic form has a short-term selective advantage.

Which explanation is correct? The case for the 'balance' hypothesis is a strong one, particularly because of the common occurrence of wholly parthenogenetic strains. It is perhaps worth pointing out that a species such as *Daphnia* does have a life history which makes it fairly plausible that there would be a short-term advantage to sex, for the kind of reasons to be discussed in Chapter 6. This interpretation is strengthened by Young's evidence for strong inter-clonal selection, although one is left with the puzzling evidence that many clones can coexist.

Before leaving the cladocerans, a word must be said about recent work on the relation between predation and life histories. A single cladoceran clone may, without sex and in response to changes in the environment, produce forms with different morphologies, including the presence or absence of spines on the carapace. This phenomenon of 'cyclomorphosis' (Hutchinson, 1967) appears to be related to changes in predation pressures. For example, Kerfoot (1974) has shown that the cladoceran *Bosmina longirostris,* in a lake in Michigan, produces small eggs and adults in summer when fish predation is heaviest on the larger individuals, and larger eggs and adults in the late fall when the main predators are invertebrates, which attack smaller individuals. Here the difference in predation pressure is temporal, and is met by a developmental change. In the cyclically parthenogenetic monogonont rotifers (see below), a similar morphological change often occurs; a female can be stimulated to produce altered and more spiny young by the presence of a predator in the water.

In tropical lakes, the predators are always present, but there may be spatial variations in predator pressure. This challenge is most easily met by genetic difference between clones. Zaret (1972) describes two forms of the cladoceran *Ceriodaphnia cornuta* in a lake in Panama. In open water, the only form is hornless and has a high rate of reproduction. Near the shore, where plankton-eating fish are abundant, there is also a horned form. These forms are different clones, or groups of clones. Males are not found, perhaps because 'winter eggs' are not needed in a permanent tropical lake.

What is the significance of the sexual phase of the life history? With fairly predictable annual cycles containing many generations it is clearly better to track changes developmentally, by cyclomorphosis, than by genetic change. Mr Lester Lee has suggested to me that sex is important in producing new clones which react differently to environmental cues, because the relationship between predation pressure and such cues as temperature and photoperiod will vary from year to year.

Effectively, the same picture emerges from other cyclically parthenogenetic groups. In all cases the sexual and asexual forms play different ecological roles; usually there are strains which have lost the sexual phase.

In digenetic flukes, apomictic parthenogenesis occurs in the secondary host, usually a snail. Hermaphrodite sexual forms inhabit the final host – a vertebrate.

Among the rotifers (Birky & Gilbert, 1971) one order, the Monogononta, have a life history astonishingly similar to that of *Daphnia*. Like *Daphnia,* they inhabit fresh water. All-female populations reproducing parthenogenetically, probably by apomixis, can persist indefinitely. The parthenogenetic eggs develop immediately. Occasionally, under the influence of high density, sexual females are produced parthenogenetically. If these females are not mated (during the first few hours of life) they lay unfertilised eggs which develop into degenerate haploid males; if they are mated, they lay diploid 'winter eggs' which play the same ecological role as winter eggs in *Daphnia.* That is, they do not hatch immediately, but ultimately hatch into a new strain of parthenogenetic females. It is not uncommon in the Monogononta for sex to be lost altogether; as one would expect, it happens most commonly in lakes where the need for a resistant egg is less. However, at least one case has been reported (Ruttner-Kolisko, 1946, quoted by Gilbert, 1974) in which winter eggs were produced parthenogenetically, without males.

The gall wasps (Cynipidae) typically have two generations only in the year, a parthenogenetic one in the summer and a sexual one in the winter. Again, there is an ecological difference between the types and locations of the galls produced by the two forms, and again wholly parthenogenetic strains are not uncommon.

The aphids have elaborated the mechanism of cyclical partheno-genesis to the highest level of complexity. There is not only an ecological differentiation between the sexual and parthenogenetic phases, but there are usually a number of ecologically distinct parthenogenetic forms following one another in more or less regular sequence. As in the other groups with an annual cycle, a number of wholly asexual strains are known. These are particularly common among aphids infesting greenhouse crops, in which no overwintering phase is called for. However, it seems likely that ecological forms which normally develop from fertilised eggs can develop partheno-genetically.

What can we conclude? These cyclically parthenogenetic species certainly fit well with the idea that there is a short-term balance between the advantages of sex and parthenogenesis. Further, they have a population structure which makes a short-term advantage for sex relatively easy to envisage. But the ecological differentiation between sexual and asexual phases makes it impossible to dismiss the group selection explanation entirely.

Facultative parthenogenesis

If a natural population is found to contain an excess of females it is essential to distinguish between two possibilities. The population may contain individual females capable either of reproducing apomictically or, after fertilisation, of producing male and female offspring sexually. If so, the sexually and asexually produced offspring will have the same range of genotypes and their coexistence in nature proves that there is a short-term advantage for sexual reproduction. Alternatively, the population may contain a mixture of obligate parthenogens and normal sexual males and females. If so, the parthenogenetic part of the population may contain a restricted range of genotypes, and this may account for its failure to replace its sexual competitors (section F, below).

In at least a few plant species, it seems clear that there is a short-term balance between sex and parthenogenesis. An example is the grass *Dichanthium aristatum* (Knox, 1967). Individual plants produce haploid egg cells which require fertilisation in the normal way and diploid embryos genetically identical to the parent plant. Knox found that the proportion of apomictic offspring varied from 60 to 90%, according to the day-length regime to which the plants were exposed. It is a safe general rule that if phenotypic expression can be altered by environmental means, there will be genetic variance as well. *D. aristatum* provides clear support for Williams's balance argument.

The situation is less clear in animals. Cases in which an occasional egg develops without fertilisation are not relevant, particularly if such development is automictic. There is one group of animals, the stick insects (Phasmidae), in which it is common for the unfertilised eggs laid by females of sexual species to develop. In typically bisexual species, as many as 50% or more of unfertilised eggs may hatch (Bergerard, 1962). An example is *Clitumnus extradentatus*, where meiosis is normal and restitution is achieved by fusion of cleavage nuclei. It seems likely that the enforced homozygosity this entails would more than counterbalance the advantages of parthenogenesis, so that the species remains bisexual. However, the same type of automictic parthenogenesis occurs in *Bacillus rossius* (Pijnacker, 1969), which is entirely parthenogenetic in France but bisexual in North Africa. It follows that at least some homozygous genotypes are sufficiently fit to have replaced their sexual ancestors.

If all facultative parthenogenesis in stick insects were automictic, one could argue that the parthenogenetic varieties are normally eliminated selectively because of their homozygosity, and therefore lend no support to the balance argument. However, *Sipyloidea sipylus* has been described as apomictic (Pijnacker, 1967). This species is bisexual but facultatively parthenogenetic in the Malay Archipelago, its original home, but is exclusively thelytokous in Madagascar, where it has been introduced (White, 1973). According to Pijnacker (1967), there is a pre-meiotic doubling of the chromosome number, followed by two maturation divisions which the author interprets as equational (i.e. producing offspring genetically identical to the parent). If this interpretation is correct, the maintenance of bisexual reproduction in Malaya would imply a twofold short-term advantage for sexual reproduction.

Although the phasmids are unique among animals in that it is common for eggs laid by virgin females from typical bisexual species to hatch, facultative parthenogenesis is known (or at least strongly suspected) to occur in nature in some other cases. Two examples are described below.

Calligrapha is a genus of chrysomelid beetles, consisting of many sibling species living on different food plants. Robertson (1966) studied seventeen nominal species from North America. Of these, five are obligate tetraploid thelytokes; eleven are diploid bisexual species. One, *C. scalaris,* is an obligate tetraploid thelytoke in Kansas; in Ottawa it is a diploid and almost certainly facultative thelytoke (the evidence strongly suggests that individual females produce female offspring parthenogenetically if unmated, and male and female offspring sexually if mated; however, it does not completely rule out the possibility that natural populations are a mixture of obligate thelytokes and sexual individuals). Thelytoky is probably apomictic. The numbers of eggs laid in a season by sexual and parthenogenetic females are approximately equal. In the majority of the sexual species there is a substantial excess of females in natural populations, but it is not clear that this indicates facultative thelytoky.

A very similar situation exists in the hydrobiid snail *Potamopyrgus* (Winterbourne, 1970). There are three species in New Zealand, of which one, *P. antipodorum,* is thelytokous and the other two bisexual; all three are diploid. Parthenogenesis is thought to be ameiotic. *P. antipodorum* is abundant and morphologically variable. Most populations consist of females only, but in some functional males exist also, usually in a minority. It is thought that females are facultatively thelytokous, producing male and female offspring sexually when mated but, as in the case of *Calligrapha,* the possibility that the species consist of a mixture of sexual individuals and obligate thelytokes has not been ruled out. (Incidentally, it is very likely that *P. jenkinsi,* a wholly apomictic species introduced into Europe during the last century and now very abundant, is derived from *P. antipodorum.*) Of the two sexual species, one has a 1:1 sex ratio in natural populations and the other an excess of females, but, again as in *Calligrapha,* it has not been demonstrated that thelytoky is responsible for the disturbed sex ratio.

Thus in both *P. antipodorum* and *C. scelaris,* there are populations which are in all probability facultatively thelytokous. There is also an indication that the balance between thelytokous and sexual reproduction may be ancient, since there are related bisexual species with an excess of females in natural populations. Further investigation of these cases could be most interesting.

Pseudogamy

The genera *Poa* and *Potentilla* (Stebbins, 1950, 1971; Muntzing, 1958) contain complex arrays of obligate and facultative apomicts, and a few sexual forms. Facultative apomicts are found throughout the geographic range of both species. In both genera apomictic plants produce functional pollen, and usually require fertilisation before they set seed, although some apomictic varieties of *Potentilla* are capable of autonomous reproduction without fertilisation. The function of the pollen in apomictic reproduction is to fertilise the endosperm.

There is an important difference between the evolution of these groups and a group such as *Taraxacum*. In the latter, new obligate apomicts are produced from the sexual species, which have a restricted range; in *Poa* and *Potentilla*, new varieties are produced from facultative apomicts. The long-term future of *Taraxacum* may be, first, the extinction of the sexual forms in competition with their apomictic descendants, and then the extinction of the apomicts in competition with unrelated sexual species which have retained greater evolutionary potential. *Houttuynia cordata*, an obligate apomict without sexual relatives, may represent a late stage in this process (Babcock & Stebbins, 1938). In contrast, the widespread occurrence of facultative apomixis in *Poa* and *Potentilla* ensures the continued production of new genetic varieties.

There are three questions concerning the selective forces acting on pseudogamous complexes such as *Poa* and *Potentilla:*

Why do apomictic plants continue to produce functional pollen? At first sight, it might seem that since no genes are transmitted to future generations by a pollen nucleus which fertilises the endosperm, a mutant gene suppressing pollen production would be favoured.

However, this overlooks two points. First, genes may be transmitted to future generations by a pollen nucleus which fertilises the egg nucleus of a facultative apomict. Secondly, even if we consider selection between clones of obligate apomicts, there will still be selection for producing pollen, so long as clones continue to require fertilisation before they set seed. This arises because individual plants will tend to grow close to their relatives – i.e. to members of their own clone. Hence by producing pollen a plant is helping to ensure the fertility of plants genetically identical to itself; pollen production will be favoured by kin selection.

Why do apomictic plants require fertilisation? This question is much harder to answer. In *Potentilla* there are varieties capable of autonomous reproduction. Such a change would seem to confer some advantage, at least in colonisation, without any obvious corresponding disadvantage. A group selectionist explanation is possible: varieties which become autonomous will tend to lose the capacity to produce functional pollen, and so will become genetically isolated obligate apomicts which are doomed to extinction. I cannot pretend to feel much enthusiasm for this explanation.

Why do facultative apomicts continue to produce sexual as well as asexual progeny? This is, in the present context, the most relevant question. There is plenty of evidence (e.g. Muntzing, 1958) that there are genes which influence the proportions of the two types of offspring produced. It is also easy to see that, in the absence of any short-term selection in favour of sexual reproduction, an allele causing apomixis will increase in frequency relative to an allele causing meiosis. This seems to be a clear case where Williams's (1975) 'balance' argument applies.

F   What is it that goes extinct?

If one is to use the balance argument to prove that there is a short-term selective advantage to sexual reproduction, it is necessary to show that there are sexual populations which have the genetic capacity to evolve into obligate thelytokes but which do not do so. It is also necessary that the thelytokes should not be handicapped by having excessively homozygous genotypes. In effect, therefore, one is seeking cases of facultative apomictic parthenogenesis, or of cyclical parthenogenesis in which there is no ecologically specialised phase which can be produced only by sexual means. Some such cases do exist (e.g. in *Dichanthium, Sipyloidea, Potentilla, Daphnia,* and perhaps *Calligrapha* and *Potamopyrgus*), and they may be commoner than is at present apparent.

Much more commonly, however, we have evidence of the coexistence of a bisexual species and of one or more parthenogenetic varieties derived from it. Indeed, this state of affairs is so common as to be the rule rather than the exception. There is no reason to think that this state of affairs is transient in the short run. What it means is that a parthenogenetic clone can often establish itself in competition with its sexual ancestor, but cannot wholly eliminate the latter.

This is not unexpected. The origin of any particular parthenogenetic clone is a unique event, and all members of that clone will initially have the same genotype (see, e.g. Cuellar, 1976). If this is a reasonably well-adapted genotype, the intrinsic twofold advantage of parthenogenesis will enable the clone to exclude the sexual population from part of its geographical or ecological range. But if the parental species has a reasonably wide ecological niche, a single clone is unlikely to be able to displace it completely, even given a twofold advantage. The result will be the coexistence of the parthenogenetic and sexual forms.

It is interesting in this context that in the genus *Calligrapha,* which consists of a series of sibling species adapted to single food plants, the species are, with one exception, either sexual or parthenogenetic, suggesting that a single clone may have been sufficient to eliminate a sexual species. In contrast, a plant such as *Taraxacum* may find itself in a wide range of different situations; a large number of apomictic biotypes can coexist, and have not yet completely eliminated the sexual forms.

If sexual and parthenogenetic forms do coexist, one of two outcomes is possible. On the one hand, the sexual forms may give rise continuously to new parthenogenetic varieties, with the result that the whole ecological range of the sexual species comes to be occupied by apomicts, with the consequent extinction of the sexual species.

Alternatively, the greater evolutionary potential of the sexual population may enable it gradually to exclude the parthenogenetic clone or clones. The result will depend on the relative rates of the two processes, the generation of new parthenogenetic varieties and the evolutionary improvement of the sexual variety.

The relevance of this to group selection is as follows. A single parthenogenetic mutant, even if it does establish itself, will not normally replace a whole species; usually it merely excludes the parental species from part of its range. It follows that the requirement for group selection to operate, *i.e.* that each parthenogenetic mutant must be balanced by a species extinction, is too severe. The group selection explanation is to that extent more plausible.

Moore (1976) has attempted to measure the fitness of a parthenogenetic variety, with illuminating results. *Poeciliopsis monacha-occidentalis* is a diploid parthenogenetic fish which is an F_1 hybrid between the two sexual species, *P. monacha* and *P. occidentalis* (see p. 47). The fish are all females, and carry a complete genome from each parental species. In meiosis the paternal genome, from *P. occidentalis,* is eliminated. The fish must mate with a male *P. occidentalis,* who provides a new *occidentalis* genome to replace the one which has just been eliminated in the polar body. The parthenogenetic variety is therefore found only in regions where *P. occidentalis* is also present.

The total fitness of the hybrid form, relative to the sexual one, comprises three components:

(i) a twofold advantage, because the hybrids produce only daughters;

(ii) a mating disadvantage, because *P. occidentalis* males prefer to mate with females of their own species (as well they might);

(iii) a 'primary' fitness, including differences in fecundity and survival.

Moore estimated empirically the disadvantage suffered by the hybrids because of mating preferences. As one would expect, this disadvantage is more severe when the hybrid is common relative to the sexual form, and is virtually absent when the hybrid is rare. In consequence, there is a frequency dependence which can maintain both types in balance in a natural population. If there were no differences in 'primary' fitness, this balance would be at 85% hybrids to 15% sexuals. Over most of the range the hybrid is present in lower frequency than this, suggesting that its primary fitness is lower than that of the sexual, so that it is maintained only by the twofold advantage. In some regions occupied by *P. occidentalis* the hybrid is wholly absent. The hybrid is present at higher than 85% frequency only at the southern boundary of the distribution of *P. occidentalis*, where it overlaps the range of *P. monacha*.

This suggests that the F_1 hybrid genotype is indeed fitter than that of the parental species, but only in a narrow ecological region where conditions are intermediate between those to which the parents are adapted. Elsewhere, the hybrid is present only because of its twofold advantage. The case is a special one, because of the unique genetic situation. However, it does show that a single hybrid genotype may be able to outcompete a varied sexual population over part of the latter's range, but be unable to replace it altogether.

G   Conclusions

We cannot return a simple yes or no answer to the question whether sex is maintained by group selection. It seems clear that group extinction does play some role. The taxonomic distribution of apomictic groups, when contrasted with groups having other reproductive adaptations, makes it certain that a group which does wholly abandon sexual reproduction has a limited evolutionary future. The only organisms which challenge this conclusion are the bdelloid rotifers, which stand in such sharp contrast to other parthenogens that they constitute a real problem for evolution theory. Despite the bdelloids, the contrast is so clear that one can turn Williams's (1975) extinction argument on its head, and assert that the taxonomic distribution of parthenogens demonstrates that such populations go extinct because they lack evolutionary potential, not because they run out of niche.

However, to show that wholly parthenogenetic groups are usually doomed to extinction is not the same as showing that group selection is the major selective force maintaining sex. It is quite possible that of every one hundred parthenogenetic mutants that arise, ninety-nine are eliminated because they suffer a short-term disadvantage in competition with sexual forms, and only one is established. If this one were ultimately eliminated because of lack of evolutionary capacity, it would remain true that the major selective force maintaining sex is short term.

The case for regarding group selection as the major selective force depends on the proposition that adequate parthenogenetic varieties arise very rarely. By adequate, I mean that the variety should have a genotype conferring a fitness at least as high as typical members of the sexual population from which it arose, a restriction which rules out most cases of automictic parthenogenesis. It is difficult to see how one could test this proposition, but it is not obviously false. It is perhaps more likely to be true of higher animals than of plants; a single cell of a higher plant can often give rise, without meiosis, to a new plant, whereas in higher animals the differentiation of the germ line is more rigid, and the onset of embryogenesis seems to be intimately linked to meiosis. Consequently, ameiotic parthenogenesis seems to be much commoner in plants than in animals.

The strongest evidence against the group selection hypothesis comes from species with facultative parthenogenesis (e.g. *Dichanthium, Poa, Potentilla*), or with cyclical parthenogenesis, if it can be shown that there is no ecologically specialised stage which can only be produced by sexual means (e.g. *Daphnia*). In these cases there must be some short-term balance between the advantages of sex and parthenogenesis.

The origin of a parthenogenetic variety is usually a unique event, giving rise to a genetically uniform clone, although this clone may acquire genetic variation later by mutation or by hybridisation. Such a clone may displace its sexual progenitors from part of their ecological or geographical range but, except for species with a narrow ecological niche, it is unlikely to displace them altogether. It follows that, in the balance between parthenogenetic mutants and species extinction, many successful mutants will usually be required in order to cause an extinction.

The picture which emerges, then, is one in which both group extinction and short-term selective forces play a role in maintaining sexual reproduction. It still remains to discuss what the short-term forces are, but they must be strong if they are to overcome the cost of producing males. Finally, even if we accept that group selection has played a significant role in maintaining sexual reproduction, it cannot have contributed importantly to the origins of sex; nor can it be important in determining chiasma frequency and recombination rates, for which within-population genetic variance is widespread.

5 Recombination – the problem

Preamble

The balance argument that sex must confer a short-term advantage is persuasive, but is weakened by the fact that there are relatively few facultatively parthenogenetic populations. In this chapter, I discuss the balance argument as it applies to recombination frequency in a sexual population. The basic assumption is that, for whatever reason, the population is committed to sexual reproduction. The problem is how the rate of recombination between loci is determined.

In section A, I give the evidence that there is within-population genetic variance for the frequency of recombination. Although there are many unanswered questions, there can be little doubt that such genetic variance exists, and there is a strong presumption that it exists in almost all sexually reproducing populations. If so, the actual recombination frequency presumably represents a balance between forces tending to increase and to decrease it. This conclusion is strengthened by the argument in section B that there are forces tending to decrease recombination; although the argument is mainly theoretical, some observational evidence is available to support it (section C). If there are forces tending to decrease recombination frequency, there must also be forces tending to increase it; if there is within-population genetic variance, the forces must act over a relatively short time scale.

What these short-term forces may be is discussed in Chapters 6 and 7. There is, however, one possible short-term force which it is convenient to discuss immediately. It may be that chiasma formation, and hence genetic recombination, is selectively favoured because without it meiosis would fail; that is, meiosis would produce aneuploid gametes. It is certainly true that an effective meiosis requires that homologous chromosomes should pair, and pairing is usually accompanied by chiasmata. Therefore, it may be that selection favours the formation of chiasmata because in their absence aneuploid gametes are produced, and not because of any feature of the genetic constitution of the offspring.

Felsenstein (1974) has called such selective forces 'extrinsic', as opposed to those which are 'intrinsic' to the genetic processes themselves. Unfortunately it is hard to remember whether extrinsic means 'extrinsic to meiosis' or 'extrinsic to population genetics'. I prefer to call selection ensuring the proper functioning of meiosis 'physiological'; selection acting on the genetic consequences of recombination will be called 'genetic'.

It seems most unlikely that the short-term advantages of chiasmata are solely physiological. In male Diptera an effective meiosis is achieved without chiasmata; it is hard to believe that females could not dispense with chiasmata as well, if it were selectively advantageous to do so. More generally, even if chiasmata are necessary in most species for proper chromosome disjunction, it seems certain that they would be located in terminal regions of the chromosome if genetic recombination were selectively disadvantageous.

If the short-term forces favouring reduced recombination were strong enough, we would expect to find in natural populations a large degree of linkage disequilibrium (i.e. non-random assortment of alleles on chromosomes). Section C reviews the evidence for linkage disequilibrium in natural populations. So far as it goes, this evidence suggests that there is little linkage disequilibrium in nature; if so, the short-term forces favouring reduced recombination may be relatively weak. Evidence is presented to show that such forces do nevertheless exist.

It would be dangerous, however, to suppose that natural populations are in complete linkage equilibrium. Two important exceptions are discussed in section D: inversions and 'supergenes'. These exceptions are interesting and important, but they should not be taken as the rule.

A   Is there genetic variance for recombination frequency?

For obvious reasons, we have less information about this topic than we would like. It is easy to measure wing length or to count bristle number; in order to measure the recombination frequency of a single individual, one must raise a family, and classify the offspring for at least two phenotypic characters. Even then, one has only measured recombination frequency in one region. To select for chiasma frequency per genome, Shaw (1971, 1972, 1974) took a testis biopsy for each male and counted chiasmata. It is not surprising that information is sparse.

I will be concerned here with genes affecting recombination frequency, rather than with structural changes in chromosomes, since most within-population variance is likely to be genetic rather than structural.

Fortunately, the general picture is clear (Table 1). With few exceptions, those who have attempted to alter recombination frequency by selection have succeeded. Those who have achieved progress under selection in at least one direction include: in *Drosophila melanogaster*, Detlefsen & Roberts (1921); Chinnici (1971a); Kidwell (1972a,b); Abdullah & Charlesworth (1974): in *Tribolium*, Dewees (1970): in the lima bean, Allard (1963): in *Schistocerca gregaria*, Shaw (1972). Of these, Shaw selected for total chiasma frequency; Kidwell (1972b) and Abdullah & Charlesworth (1974) selected those individuals which were non-recombinant for a particular pair of markers; the remaining experiments depended on selecting non-recombinant individuals from those families showing the lowest recombination frequency between a particular pair of alleles. The only wholly negative result was obtained by Acton (1961).

In all these cases (with the possible exception of the early experiment by Detlefsen & Roberts (1921)), it is reasonably certain that the changes depended on genes influencing chiasma frequency, and not on changes in chromosome structure (e.g. inversions). It is true that only four species are involved, but in the absence of negative results from other species, the presumption is strong that genetic variance for recombination frequency will be found whenever it is looked for. There remain, however, a number of questions to which we would like to know the answer. Some of these are now considered, more in the hope of stimulating further research than of supplying definitive answers.

Table 1. *Selection experiments on recombination*

Organism	Mode of selection	Response	Reference
Drosophila melanogaster	Family. Up and down	Down only	Detlefsen & Roberts (1921)
D. melanogaster	Family. Up and down	Up and down	Chinnici (1971*a*)
D. melanogaster	Family. Up and down	Up only. High rec. alleles recessive	Kidwell (1972*a*)
D. melanogaster	Individual. Down only	Response in 4/6 lines. High rec. alleles recessive	Kidwell (1972*b*)
D. melanogaster	Individual. Down only	Response in 2/2 lines	Abdullah & Charlesworth (1974)
Schistocerca gregaria	Male chiasma frequency. Up and down	Up and down. Genetic variance additive	Shaw (1972)
Tribolium castaneum	Family. Up and down	Up in 2/2 lines. Down in 1/2 lines	Dewees (1970)
Lima bean	Family. Up and down	Up only	Allard (1963)

rec. = Recombination.

Are the effects local or general?

Do particular genes alter recombination in some specific region of the genome, or do they have a general effect? If their effects are specific, are they confined to the region in which the gene itself is situated? Of the experiments mentioned above, only Chinnici's (1971*b*) provides much information on these points. He selected for changes in recombination frequency between the sex-linked genes *cv* and *sc*. Chromosome substitution studies showed that the effects produced were very specific to that region, but that the genes responsible for the effect were present on all the major chromosomes, X, 2 and 3.

The other investigations add little. Shaw (1972) selected for a change in the total number of chiasmata, so that the responses he obtained must have been the result of genes with general effects, or, less probably, of a large number of genes each with specific effects. He does point out, however, that changes in total chiasma frequency are accompanied by changes in chiasma location. Kidwell (1972*b*) did not attempt to locate the selected genes, but she does point out that the rapid response to selection would most easily be explained by a recombination-reducer closely linked to the selected region.

Thus, what little information we have suggests that genes may be specific or general in their effects, and, if specific, they may affect their own chromosome region or some distant region, even on another chromosome.

What is the nature of the genetic variance?

Are alleles reducing recombination more commonly dominant or recessive? Is it easier to increase or to decrease recombination by artificial selection? The second of these questions is interesting for the following reason. If selection generally acts to reduce recombination in nature, then it would be easier to increase it by artificial selection, and vice versa; this is a particular example of a point made by Falconer (1960).

Some relevant data are presented in Table 1. With the exception of the early experiment by Detlefsen & Roberts (1921), in which the downward response may have been due to structural heterozygosity, it has proved somewhat easier to increase than to decrease chiasma frequency, and there is some tendency for alleles for low recombination to be dominant; however, these effects are by no means regular enough to justify any strong conclusions concerning the direction of natural selection.

Perhaps the most informative studies on genetic variation in recombination are those of Catcheside (1968, 1975) on *Neurospora crassa*. Catcheside reports detailed studies on three mutants, located on three different chromosomes, affecting recombination rate. In all three cases the allele increasing recombination rate is recessive. All three are known to affect recombination at several locations, including sites on a chromosome other than that on which they are themselves situated. Effects of two kinds have been measured – intracistronic recombination (by seeking wild-type recombinants between different mutant forms of a single gene) and recombination between different gene loci. Two of the alleles are known to have effects of both kinds. It is not known whether any of the mutants influence recombination in the region of chromosome where they are themselves situated.

B   Selection against recombination in a uniform environment

The theme of this section is that when a population has reached genetic equilibrium in a uniform environment, selection will favour a reduction in genetic recombination. The basic idea was first put forward by Fisher (1930) when he argued that, if polymorphic alleles at two loci interact in their effects on fitness, selection will tend to reduce recombination between them. Unfortunately, he gave no proof of his assertion. It is possible, however, to see qualitatively why he reached his conclusion.

The argument is easiest to see for a sexual haploid population, but can be extended to a diploid one. Suppose the population to be polymorphic at two loci, say *A,a* and *B,b*, and that genotypes *AB* and *ab* are of high fitness and *Ab* and *aB* of low fitness (the polymorphism would have to be maintained by some kind of frequency-dependent selection). Then most individuals surviving to breed would be either *AB* or *ab*. Recombination would therefore be relevant mainly in the genotype AB/ab and would give rise chiefly to offspring of low fitness. More generally, one can simply say that since surviving individuals have genotypes of higher than average fitness, they only stand to lose, in a uniform environment, by producing recombinant offspring.

A serious mathematical attack on this problem was started by Kimura (1956) and Lewontin & Kojima (1960), and has been developed further by Bodmer & Felsenstein (1967), Nei (1967), Turner (1967), Franklin & Lewontin (1970), Karlin & Feldman (1970), Lewontin (1971), Karlin & McGregor (1972), Feldman, Franklin & Thomson (1974) and others. It is not my aim to give a detailed account of this development, but to extract from it some of the main conclusions in an easily understood form. In doing so, I have been greatly helped by the treatment in Lewontin (1974).

It is convenient to follow Lewontin's notation, in which genotypes are described in the binary notation, so that the genotypes at a locus in a diploid are 11, 10 and 00; in the two-locus, two-allele case, the four possible gamete types are 11, 10, 01 and 00, and the coupling and repulsion double heterozygotes are 11/00 and 10/01, and so on. The frequencies of the four gamete types in the two-locus case are $g_{11}, g_{10}, g_{01}, g_{00}$, and the 'coefficient of linkage disequilibrium' (Lewontin & Kojima, 1960; see also p. 14) is

$$D = g_{11}g_{00} - g_{10}g_{01}.$$

The first important point to establish is that in the absence of selection (and, as is assumed throughout this section, in an infinite population), D rapidly approaches zero. If R is the recombination fraction between two loci, then after t generations,

$$D_t = (1 - R)^t D_0, \tag{5.1}$$

or, for small R,

$$D_t = e^{-Rt} D_0. \tag{5.2}$$

Thus with 1% recombination ($R = 0.01$), 500 generations are sufficient to reduce D to $1/150$ of its initial value.

In the presence of selection, it is possible that D does not equal zero at a selective equilibrium. It is convenient to discuss the doubly symmetric fitness scheme:

Locus A

		11	10	00
	11	a	b	a
Locus B	10	c	d	c
	00	a	b	a

This system will come to a stable polymorphic equilibrium provided that there is overdominance; that is, provided $d > a$, and $|d - a| > |b - c|$.

Given that there is a stable equilibrium, it can be of two kinds:

(i) $D = 0$, and all gametic frequencies are equal; this is the equilibrium one expects in the absence of linkage, or

(ii) $D \neq 0$; g_{00} and $g_{11} > \frac{1}{4}$; g_{10} and $g_{01} < \frac{1}{4}$,

or g_{00} and $g_{11} < \frac{1}{4}$; g_{10} and $g_{01} > \frac{1}{4}$.

Which type exists depends on the fitnesses and on R. If the fitnesses are additive between loci, then $D = 0$. By additive is meant that the gain of fitness by being doubly heterozygous, $d - a$, is the sum of the gains of fitness by being singly heterozygous at the two loci, $(b - a) + (c - a)$. Thus perfect additivity implies that $d - a = b - a + c - a$, or $a + d - b - c = 0$. Given perfect additivity, the stable equilibrium is that with $D = 0$. A measure of the degree of non-additivity is given by

$$\varepsilon = a + d - b - c. \tag{5.3}$$

If equilibria with $D \neq 0$ are to be stable, it is required that $\varepsilon > 0$; that is, the fitness of the double heterozygote must be greater than would be predicted from the sum of the two single heterozygotes.

For any given value of ε, there is a critical value of R, given by

$$R = \varepsilon / 4d \tag{5.4}$$

such that if R is less than this value, the stable equilibrium is for $D \neq 0$, and if R is greater, the stable equilibrium is for $D = 0$. Thus it is not true that any departure of the fitnesses from additivity ensures that there will be linkage disequilibrium; there will be linkage disequilibrium only if the loci are sufficiently tightly linked. Some idea of what is meant by 'sufficiently tight' can be gained from Table 2.

Table 2. *Critical recombination values, R, for linkage disequilibrium with various fitness schemes*

	Fitnesses			
Selection scheme	Double homozygote a	Single heterozygote $b = c$	Double heterozygote d	R
Additive	0.8	0.9	1.0	0
Multiplicative	$\begin{cases} 0.81 \\ 0.9801 \end{cases}$	0.9 0.99	1.0 1.0	0.0025 2.5×10^{-5}
Extreme non-additivity	0.9 0.99	0.9 0.99	1.0 1.0	0.025 0.0025

Multiplicative fitnesses are what we expect if selection acts independently on the two loci. Although formally non-additive, they are not particularly favourable for the maintenance of linkage disequilibrium, basically because, when selective advantages are small, additive and multiplicative schemes are almost identical (that is, $(1 + s)(1 + t) \simeq 1 + s + t$). Even with selective advantages of the order of 10%, linkage disequilibrium requires that the loci should show less than 0.25% recombination.

It is easy to see that when the stable equilibrium is for $D \neq 0$, the mean fitness of the population is increased by the presence of linkage disequilibrium. For example, when $a = b = c = 0$ and $d = 1.0$, linkage disequilibrium will increase the frequency of high-fitness double heterozygotes, $11/00$ (or $10/01$ if the population is at the repulsion equilibrium). Unfortunately, the reverse statement, that if linkage disequilibrium would increase the mean fitness then selection will produce it, is not true; the existence of a critical value of R means that populations do not always reach their maximum fitness.

One is strongly tempted to make the further statement that, whenever there is linkage disequilibrium in a population at a stable genetic equilibrium in a uniform environment then there will be selection for reduced recombination. This statement has not been proved in the general case, but it is known to be true in many particular cases and not known to be false in any. We are unlikely to be seriously misled if we assume it to be true when thinking about the evolution of recombination. At least it accords with the idea that an individual which has survived in a given environment can only lose by recombination if its offspring will meet the same environment. Before we can evaluate the strength of such selection, we need to know how often linkage disequilibrium exists in natural populations. The observational evidence is discussed in the next section; I want here to consider some theoretical issues. The estimates in Table 2 suggest that it may be rather unusual for linkage to be tight enough, and selection strong enough, to produce linkage disequilibrium. The 'supergenes' discussed on p. 85 would then be peculiar, with the rest of the genome in linkage equilibrium, rather than being the tip of an iceberg of linkage disequilibrium.

Two processes have been suggested which might maintain a large amount of disequilibrium in natural populations, without requiring unreasonably large selective pressures. The first is 'normalising selection'; that is, selection favouring individuals close to the mean of the phenotypic distribution at the expense of the extremes. Thus suppose there is a series of + and − alleles. The optimum phenotype could be produced by a homozygous genotype, for example + − + −/+ − + −, or by multiple heterozygotes such as + − + −/− + − +. Mather (1943) was the first to consider such genetic systems. He argued that populations containing balanced genotypes of the latter kind would be favoured because they would combine high fitness in the contemporary environment with an ability to release variability by recombination if the environment changed. The argument appears to be group selectionist, but Mather may have been right for the wrong reasons.

Lewontin (1974) has re-examined the problem from a different viewpoint, using computer simulation of a five-locus model, with selection for an intermediate optimum. He found that with free recombination between loci, the population progressed rather rapidly towards homozygosity, being effectively homozygous at four of the five loci after ninety generations (since the optimum phenotype required five + and five − alleles, the population could never become homozygous at the fifth locus). In contrast, if the loci were tightly linked, there was only a slight reduction in heterozygosity after ninety generations. Instead, the population contained high frequencies of complementary repulsion gametes, for example $+ - + - +$ and $- + - + -$. Ultimately, the population would go to homozygosity, but very slowly. It follows that actual populations under normalising selection may exhibit a high degree of linkage disequilibrium, not because this is a stable state, or because of any long-term group selection, but because the approach to the genetically homozygous equilibrium is so slow.

Although the approach to genetic homozygosity in Lewontin's simulations was greatly slowed down by linkage, the initial rate of increase in mean fitness of the population was greater when linkage was tight. This illustrates Eshel & Feldman's (1970) finding that recombination can slow down evolutionary progress when there are fitness interactions between loci. It is important that selection for an intermediate optimum inevitably causes strong fitness interactions between loci; $+ - / + -$ is fitter than either $+ + / + +$ or $- - / - -$.

The second reason for thinking that linkage disequilibrium is more likely to arise in natural populations than is implied by the values in Table 2 is that the two-locus model may give a misleading picture of what happens when many loci are segregating. The first analytical treatment of a three-locus system with selection and linkage was by Feldman *et al.* (1974). As in the two-locus case, there is a critical value of R above which the only stable equilibrium is with $D = 0$. A novel behaviour, which is not manifested by the two-locus model, is that for lower values of R there may be stable equilibria with $D = 0$ and $D \neq 0$, so that the final value of D depends on the initial conditions. In contrast, for the two-locus case either the equilibrium with $D = 0$ is stable (large R), or equilibria with $D \neq 0$ are stable (small R) (Lewontin & Kojima, 1960; Karlin & Feldman, 1970). The analysis of the three-locus model suggests that, for given fitnesses and selection coefficients, there may be a great number of different possible stable equilibria.

Systems with more than three loci have been studied only by simulations. Franklin & Lewontin (1970) simulated a thirty-six-locus model, with 10% heterosis at each locus and multiplicative fitnesses. For a two-locus model, this would give a critical value of R of 0.0025

(Table 2). When the loci were assumed to be linearly arranged with a recombination fraction of R between adjacent loci, the critical value of R for linkage disequilibrium was 0.01, or four times as great. For R substantially less than 0.01, 85% of the gametes in the population belonged to one of two reciprocal types. A population with such an extreme degree of linkage disequilibrium has a far higher mean fitness than one with $D = 0$.

The theoretical position can be summarised as follows. In the absence of selection, an infinite population approaches linkage equilibrium rapidly (equation 5.1). Selection in a constant environment (i.e. fitnesses constant) may generate linkage disequilibrium. If fitnesses are additive between loci (i.e. no epistasis), then $D = 0$ at a stable equilibrium. If there are epistatic fitness interactions, there may be a stable equilibrium with $D \neq 0$. However, the maintenance of linkage disequilibrium depends on the recombination fraction R being less than a critical value (equation 5.4). The relevance of linkage disequilibrium in the present context is that, if there is a stable equilibrium with $D \neq 0$, then the mean fitness of a population will be higher than it would be with free recombination, and there will be selection favouring a reduction in recombination.

The critical value of R for the two-locus case suggests that in nature linkage disequilibrium (and hence selection for reduced recombination) may be confined to a few special cases with intense selection and tight linkage. There are, however, two reasons why linkage disequilibrium may be a more widespread phenomenon. First, selection for an intermediate optimum, although ultimately leading to genetic homozygosity, may in the presence of linkage maintain a large amount of linkage disequilibrium for a long time before homozygosity is reached. Secondly, linkage disequilibrium may be maintained in multi-locus systems with smaller selective coefficients, or larger recombination fractions.

We now turn to the observational evidence.

C   Linkage disequilibrium and recombinational load – the observations

The only way at present available of deciding whether there is substantial linkage disequilibrium in natural populations is to study allozymes at linked loci, either in gametes extracted from natural populations or, less directly, in zygotes. The data so far collected on *Drosophila melanogaster* have recently been reviewed by Langley (1977). His basic conclusion is that there is little detectable disequilibrium between allozyme loci. In so far as any has been detected, it is of the kind to be expected by chance in a finite population, without selection. There is non-random association between inversions and enzyme loci included within them. This is to be expected on historical grounds, since an inversion when it first arises carries a particular set of alleles.

What is true of *D. melanogaster* seems to be true of other *Drosophila* species which have been studied. These results make it very unlikely that there is the extreme degree of linkage disequilibrium in natural populations of sexually outcrossing species suggested by Lewontin (1974). However, they do not prove that disequilibrium occurs only between very tightly linked loci. For one thing, most sample sizes have inevitably been rather small. More important, Langley points out that an electrophoretically identified category may include a number of different alleles; if so, by lumping them together we may be missing significant disequilibrium. The labour of identifying alleles uniquely in a sufficiently large sample is formidable, but may be unavoidable.

An alternative way of deciding whether selectively maintained linkage disequilibrium is important is to measure directly the loss of fitness which results when chromosomes which have been exposed to selection in a natural population are allowed to recombine. If no loss of fitness results from recombination, then there is no selectively relevant linkage disequilibrium. This approach was taken by Charlesworth & Charlesworth (1976a), using *D. melanogaster,* and relying on the special techniques available in this species to 'construct' flies with particular chromosomal constitutions. They compared components of fitness (viability, female fecundity) of flies carrying various combinations of 'M' and 'F' chromosomes; an M chromosome was extracted from a wild-caught adult male without opportunity of recombination, and an F chromosome from a wild-caught female, permitting a single round of recombination since being selected in nature.

Their results are suggestive. Measures of viability (egg to adult) showed that M/M flies were superior to M/F, which in turn were superior to F/F, but the differences were small ($\simeq 1\%$) and not statistically significant. Rather unexpectedly, a comparison of fecundities (number of eggs laid × egg hatchability) showed that M/Cy females (i.e. females heterozygous for an M chromosome and a standard marked chromosome) were superior to F/Cy females, the difference being of the order of 7%, and statistically significant.

The chromosomes used in this study were derived from a single large natural population. The experiment shows that there is selection for reduced recombination, acting through female fecundity and probably through viability. A similar conclusion emerges from the work of Mukai & Yamaguchi (1974).

The observational evidence, limited as it is, suggests that natural populations of *Drosophila* are not composed of chromosomes carrying special co-adapted sets of genes in extreme linkage disequilibrium with one another. But there probably is some degree of selectively important linkage disequilibrium. If these results prove to be typical of outbreeding species, and at present there is no reason to doubt that this is so, then the maintenance of recombination does present a problem similar to the maintenance of sex. The cost of recombination is not two-fold, but it does exist. We have to seek for counterbalancing selective forces favouring increased recombination.

D Supergenes and inversions

Females of several species of swallowtail butterfly obtain some protection from predators through their resemblance to one of several distasteful 'model' species. Clarke & Sheppard (1971) found that the mimicry patterns in *Papilio memnon* could be interpreted as being controlled by five closely linked loci, concerned with the presence or absence of tails, forewing pattern, hindwing pattern, colour of the basal triangle of the forewing, and abdomen colour. Thus the genotype of a mimic of one model species might be *AbCdE*, say, and of another be *aBcDe*. Rare intermediate forms can be interpreted as recombinants. Since in natural populations linkage disequilibrium is almost complete, and linkage is tight, the various model patterns behave almost as if they were controlled by a single gene; hence the term 'supergene' to describe the group of linked loci.

The situation corresponds precisely to that in which we would expect to find linkage disequilibrium in nature. There are strong epistatic fitness interactions, because a female which resembles one model in some features and a second model in others could easily be distinguished from both, and will therefore be of low fitness. Combined with tight linkage, this selection should lead to extreme linkage disequilibrium.

Other examples of supergenes are the loci controlling heterostyly in *Primula* (Ernst, 1936), and those controlling shell colour and pattern in snails (Murray, 1975). In *P. vulgaris,* three separate loci control style length, stamen length, and pollen characteristics. Recombination can give rise to self-compatible 'homostyles'. Such homostyles can reach a high frequency locally (Crosby, 1949), but presumably fail to replace the typical heterostyles because of inbreeding depression. In some other species of *Primula,* usually with marginal distributions, homostyly and self-compatibility are established as the rule.

It is not certain how such supergenes originate. It is most unlikely that the different loci in *Papilio* or *Primula* could have arisen by duplication, because they influence such different characteristics. These cases should therefore be sharply distinguished, at least as far as origin is concerned, from those in which duplication is a plausible origin (e.g. antigenic loci such as the HL-A system in man and the H-2 system in the mouse; see Thomson, Bodmer & Bodmer, 1976). Apart from duplication, there are three possibilities. The loci may originally have been unlinked, and have been brought together by translocations because of the selective advantage of close linkage; they may have been on the same chromosome but loosely linked, and have been brought together by inversions or by a reduction of chiasma frequency; or the loci may have been tightly linked from the beginning.

There are obvious difficulties with each of these explanations. Translocations or inversions with precisely the right break points must be rare events, and would in any case be unlikely to be established because of their effect on fertility. But it would be a fortunate chance if loci capable of mutating so as to produce the required phenotypic effects happened to be closely linked. Charlesworth & Charlesworth (1976*b*) prefer the third explanation, at least for *Papilio*, for the following reason. If selection for reduced recombination is to have a chance of bringing initially unlinked or loosely linked loci together, then there must be long-lasting polymorphism at both loci. The authors show that, for a plausible model of the selective process, a stable polymorphism at each locus would exist only if they were already tightly linked.

Whatever their origin, supergenes such as those in *Papilio* and *Primula* do confirm the theoretical prediction that with sufficiently tight linkage and with epistatic fitness interactions, extreme linkage disequilibrium can be maintained in a natural population. The same is true if recombination is prevented by inversions. In several species of *Drosophila* it is known that there is non-random association between alleles determining isozymes and chromosome order. The selective mechanisms responsible have been discussed by Charlesworth & Charlesworth (1973). It is important to distinguish between genetic differences which have arisen since the establishment of an inversion polymorphism, and those which existed before the inversion occurred and which were responsible for its establishment.

Thus, suppose there exists at some locus a pair of alleles, *A* and *a*, such that the heterozygote is fitter than either homozygote. If, in a population already polymorphic for *A* and *a*, an inversion were to occur, the new order would carry only one of the two alleles, but this would not help to establish the new order. (Of course, if the new inversion were established for other reasons, there would also be linkage disequilibrium between it and the *A* locus.) But if an inversion polymorphism existed first in an *AA* population, then when an *a* allele arose by mutation, it would occur on one order only, and linkage disequilibrium between the inversion and the gene locus would persist for a long time.

In contrast, if there is to be positive selection favouring the establishment of a new inversion, the authors show that there must already exist in the population some selectively maintained linkage disequilibrium. In effect, this conclusion follows from the fact that selection for reduced recombination occurs only in the presence of such disequilibrium. If we stand this argument on its head, the widespread occurrence of inversion polymorphisms in *Drosophila* implies the presence of some selectively maintained linkage disequilibrium in the absence of inversions.

6 *Short-term advantages for sex and recombination – I. An unpredictable environment*

Preamble

At equilibrium in a uniform environment, if there is any selection on recombination it will be for a reduction. It follows that, if there are situations in which selection favours higher recombination, they will be ones in which the environment varies in space or in time or both. In this chapter, I discuss whether there are circumstances which are likely to exist in nature, and in which there is a short-term selective advantage for sex and recombination.

Section A discusses selection in a varying environment. First, I describe in qualitative terms the kinds of environmental unpredictability which may arise, and their consequences for selection on recombination. I reach the tentative conclusion that selection will favour higher recombination only in the unlikely case that the correlation between different features of the environment changes sign from generation to generation. This conclusion is then supported by a more formal model of competition between sexual and asexual forms in a fluctuating environment, and by a simulation by Charlesworth (1976) of selection on recombination in a fluctuating environment.

The conclusion of section A is that the belief that sex and recombination are favoured in a variable or unpredictable environment is too simple. The environment must be unpredictable in a special and somewhat implausible sense. In section B, I consider a spatially varying environment with migration. There can again be selection for increased recombination, but this seems to require a rather precise relationship between the scale of environmental heterogeneity and the typical migration distance.

Section C discusses what I believe to be the essential feature of most of the models suggested by Williams (1975) as providing a short-term advantage for sex. This feature is competition between siblings or other close relatives. Qualitatively, the significance of this is as follows. A parent will leave fewer descendants if there is severe competition between its offspring. The more alike genetically its offspring are, the more intense the competition between them. A sexually reproducing parent reduces this competition by producing genetically dissimilar offspring. Sex is an advantage because it generates within-family variation.

The effects of 'sib competition' are hard to investigate analytically. I therefore describe a simulation (Maynard Smith, 1976*f*) which is intended to make more explicit the models suggested by Williams (1975) and Williams & Mitton (1973). It turns out that sib competition can indeed provide a short-term advantage for sex, or for alleles for higher recombination. However, it is far from clear that these advantages will often be sufficient to outweight the counter-balancing selection; some of the difficulties are discussed at the end of the section.

In section D I discuss the observational evidence, particularly that concerned with the geographical distribution of parthenogenetic varieties. This tends to support the view that sex is maintained in an environment in which qualitatively new selective challenges are common, rather than one in which an unchanging set of selective pressures succeed one another in an unpredictable sequence.

A   Selection in a varying environment

A qualitative account of predictability

The environment in which an individual lives can be thought of as being characterised by the 'state' of a number of 'features' – for example hot or cold, wet or dry, crowded or empty, with or without a particular predator. These features will be symbolised by barred letters, $\bar{A}$ or $\bar{a}$, $\bar{B}$ or $\bar{b}$, and so on; of course, each feature could exist in more than two states, but I do not think that this would make any essential difference to the conclusions. The environment is thought of as 'coarse-grained' in Levins's (1968) sense; that is, an individual spends its whole life in a particular environment, say $\bar{A}\bar{B}\bar{c}\bar{d}\bar{E}$. In nature animals also experience environmental changes in the course of an individual life time. This is irrelevant to our present purpose, since in a fine-grained environment selection will favour that genotype which has the highest fitness over the whole life span (Strobeck, 1975).

I at first suppose that for each feature there is a single locus with a pair of alleles, each adapted to one of the states; symbolically, allele A is fitter in environment $\bar{A}$ and allele a in $\bar{a}$, and the genotype $ABcdE$ is optimally adapted to environment $\bar{A}\bar{B}\bar{c}\bar{d}\bar{E}$. The assumption of haploidy is made for simplicity; it is dropped in the models discussed on pp. 95–6. Selection in haploid and diploid populations is, at least qualitatively, similar. A much more crucial assumption is that only a single locus is concerned with adaptation to any one environmental feature. This is discussed further on pp. 92 and 107. It turns out that if one makes the plausible assumption that alleles at several loci are concerned with adaptation to a single feature, it becomes much more difficult to find situations in which selection will favour high recombination.

We are now in a position to consider the different kinds of environmental predictability, and their selective consequences.

Range of environments predictable. Suppose that the environment varies spatially, but that the frequencies of particular states, for example $\bar{A}$ and $\bar{a}$, remain constant from generation to generation. If there are correlations between states (i.e. $\bar{A}\bar{B}$ and $\bar{a}\bar{b}$ common relative to $\bar{A}\bar{b}$ and $\bar{a}\bar{B}$), then these also remain constant. If dispersal is limited, so that offspring tend to live in the same kind of environment as their parents, then high fitness is best ensured by asexual reproduction, or mating between neighbours. If dispersal is random, then the probability that an individual of genotype G will find itself in an environment E is constant from generation to generation and is the same for all genotypes. There are then two possibilities:

(a) Fitnesses not frequency-dependent. The fitness of genotype G is then

$$\sum_{\text{all E}} P(G \text{ finds itself in E}) \times (\text{fitness of } G \text{ in E}).$$

Since both the expressions inside the summation are constant in time, each genotype has a fitness which is also constant in time. There is therefore an optimal genotype. The population will become homogeneous for this genotype. Sex makes no difference to the outcome (unless the optimal genotype is a heterozygote, in which case sex is a disadvantage).

(b) Fitnesses frequency-dependent. This is a much more likely state of affairs. If there is density-dependent population regulation within each environmental niche, genotypes will be fitter when they are less common, leading to a stable polymorphism (Levene, 1953; Maynard Smith, 1962).

In this case, if there are associations between environmental states, there will be selection against recombination. Thus if $\bar{A}\bar{B}$ and $\bar{a}\bar{b}$ are common, and $\bar{A}\bar{b}$ and $\bar{a}\bar{B}$ are rare, then genotypes AB and ab will be on average fitter than Ab and aB, and recombination will break up well-adapted genotypes.

There is a second reason why selection will favour low recombination, even if there is no association between environmental states. If alleles at two loci adapt an animal to the same environmental feature, so that genotype A_1A_2 is adapted to $\bar{A}$ and a_1a_2 to $\bar{a}$, then again recombination will tend to break up well-adapted gene combinations.

Thus, if the range of environments present is constant, but the particular environment of an individual is unpredictable, selection will either have no effect on recombination or will tend to reduce it.

Range of environment unpredictable. Suppose now that the frequencies of the possible states of the environment change from generation to generation. There are two cases to consider:

(a) Correlation between states remains constant. That is, we suppose that the relative frequencies of particular states of the environment change from generation to generation, but associations between states do not. In this case, there is either no selection on recombination or (if there are associations between states, or if more than one locus is concerned with adaptation to a single environmental feature) selection again favours reduced recombination. The

reason is as follows. A change in the frequency of a particular state can be met only by a change in gene frequency, and recombination does not alter gene frequency. Recombination does alter associations between alleles; so long as environmental associations do not alter, this lowers fitness.

(b) Correlations between environmental states change sign from generation to generation. That is, if in one generation hot places tend to be dry and cold places wet, then in the next generation hot places tend to be wet and cold ones dry. In this case, genotypes which survived in one generation have combinations of genes which are ill-adapted in the next. Selection will favour higher recombination. But surely few species can be confronted with this degree of environmental unpredictability.

The next section offers a somewhat more formal proof of these assertions (Maynard Smith, 1971*a*, but with a different notation). It can be safely skipped by anyone who accepts the conclusions just stated.

Competition between sexual and asexual populations

Consider an environment characterised by two features, each of which can be in one of two states, so that the possible environments are $\bar{A}\bar{B}$, $\bar{A}\bar{b}$, $\bar{a}\bar{B}$, and $\bar{a}\bar{b}$. Let the frequencies of these in generation n be

$$p(\bar{A}\bar{B}) = x_n y_n + d_n$$
$$p(\bar{A}\bar{b}) = x_n(1 - y_n) - d_n$$
$$p(\bar{a}\bar{B}) = (1 - x_n)y_n - d_n$$
$$p(\bar{a}\bar{b}) = (1 - x_n)(1 - y_n) + d_n. \tag{6.1}$$

The coefficient d is an obvious analogue of D, the coefficient of linkage disequilibrium (p. 78). It refers, however, to environmental characteristics instead of gamete frequencies. I suggest that it be called the 'coefficient of environmental association'. If it is zero, environmental states are uncorrelated.

We assume that one locus is concerned with adaptation to each feature, and that selection is so intense that only the optimally adapted genotype AB survives in $\bar{A}\bar{B}$, and similarly for the other habitats. Suppose that in generation n an equal number of sexual and asexual adults survive in each type of environment, the total number of each kind being T; there will then be $T(x_n y_n + d_n)$ adult sexual and adult asexual AB individuals in generation n. We ask: will there be more sexual or more asexual adult survivors in generation $n + 1$?

Each adult, sexual or asexual, produces N offspring (i.e. we ignore the advantage of not producing males); the offspring are randomly distributed between environments; an individual arriving at a patch to which it is perfectly adapted has a fixed chance C of surviving (i.e. we ignore frequency-dependent selection).

Considering first asexual individuals of genotype AB, the number surviving in generation $n + 1$ is

$$TNC(x_n y_n + d_n)(x_{n+1} y_{n+1} + d_{n+1}).$$

The total number, F_{AS}, of surviving adult asexuals in generation $n + 1$ is the sum of four terms of this kind. We now want to find F_S, the total number of surviving adult sexuals in generation $n + 1$. I assume that a single generation of sexual reproduction produces offspring in linkage equilibrium ($D = 0$). This exaggerates the effect of sex, but since we are interested only in the sign of $F_{AS} - F_S$, the exaggeration will make no difference.

The number of sexual AB offspring produced is then $TNx_n y_n$, and the number of AB survivors is therefore $TNCx_n y_n(x_{n+1} y_{n+1} + d_{n+1})$, and F_S is the sum of four terms of this kind. After a little manipulation it can be shown that

$$F_{AS} - F_S = TNC\{4d_n d_{n+1} + d_n(1 - 2x_{n+1})(1 - 2y_{n+1})\}. \tag{6.2}$$

Ignoring for the moment the second term, equation (6.2) states that the frequency of sexual individuals will increase only if d_n and d_{n+1} are of different sign. This confirms the earlier conclusions that sex and recombination are an advantage only when the correlation between environmental features changes sign.

What is the significance of the second term in equation (6.2)? If either x_n or y_n is $\frac{1}{2}$, the term is zero. Thus the term is significant only if there is environmental disassociation ($d_n \neq 0$) and if for both features there is a rare and a common state. I have investigated the competition between sexual and asexual populations with a fixed coefficient of environmental association a little further. The details are complex, but the conclusions are worth reporting.

There are two extreme cases worth considering:

Environment		$\bar{A}\bar{B}$	$\bar{A}\bar{b}$	$\bar{a}\bar{B}$	$\bar{a}\bar{b}$
Frequency	case 1	0	z	z	$1-2z$
	case 2	z	0	0	$1-z$

Thus, the cases differ in the sign of the environmental association. We ask: suppose one starts with a wholly sexual population, could an asexual population establish itself? In case 2, both clones AB and ab could invade; if both were present, the sexual population would go extinct. In case 1, only clones Ab and aB can invade. If both were present, they would increase in frequency until alleles A and B were eliminated from the sexual population, which would occupy only the $\bar{a}\bar{b}$ habitat. At this stage, obviously, an ab clone would equal the sexual individuals in fitness (remember that we are neglecting the twofold cost of producing males).

There is nothing in these conclusions to modify the assertion that d must change sign if sex is to confer an advantage.

Selection for recombination in a fluctuating environment

The analysis in the previous section, although it has the advantages of generality and simplicity, suffers from the drawback that it does not follow the frequencies of alleles for high and low recombination through a number of generations in a well-defined genetic system. A model which meets these requirements was developed by Charlesworth (1976). He considered two alleles at each of two loci, a,A, and b,B in an infinite diploid random-mating population in a fluctuating environment, and investigated whether a gene modifying the amount of recombination between them would increase in frequency if introduced into the population. His conclusions, based mainly on computer simulation, can be summarised as follows:

(a) An allele for increased recombination is favoured only if the fitnesses are such that the equilibrium linkage disequilibrium, D, between the selected loci varies in time. Although he does not prove formally that D must change sign if higher recombination is to be favoured, he was able to show such an increase only if D did in fact

change sign. There is, therefore, good agreement between his results and the analysis in the previous section.

(b) The selection pressure favouring higher recombination is vanishingly small unless the locus influencing recombination is itself closely linked to the loci affecting fitness.

(c) The equilibrium level of recombination depends on the period of the environmental fluctuation (or, in a stochastically varying environment, on the autocorrelation between successive environments), being greatest for intermediate periods of the order of three to five generations.

It is clear from this work that there can be selection for increased recombination in a fluctuating environment. But before this is taken as a general explanation of recombination, we must ponder the significance of an environment which fluctuates in such a way that the equilibrium value of the linkage disequilibrium D changes sign. This means that epistatic fitness interaction must be such that in some generations gametes AB and ab are of high marginal fitness, and in others gametes Ab and aB are of high marginal fitness. I believe that such fitness interactions will be the exception rather than the rule. The difficulty has already been discussed (pp. 91–2), and will arise again when we discuss the sib-competition model (p. 107).

B   Spatial variation of the environment

When discussing the effects of recombination on the rate of evolution (Maynard Smith, 1968a), I suggested that a typical situation in which recombination would accelerate evolution is as follows. Suppose that a population, taken as haploid for simplicity, inhabits two regions, in one of which the optimal genotype is AB and in the other ab. Suppose then that a new region, in which the optimal genotype is Ab, becomes available for colonisation. The initial population in the new region will consist of AB and ab individuals. Clearly, recombination would greatly accelerate the appearance of the optimally adapted Ab genotypes.

Unfortunately, I thought of this as a transient phenomenon, and did not ask whether in such a case it is possible to have selection for increased recombination in the short run when there is equilibrium between selection and migration. A related model has since been analysed by Slatkin (1975), who shows that linkage disequilibrium can be maintained, and that selection may favour either increased or decreased recombination. I discuss here a simplified version of Slatkin's model, which will bring out the essential features.

Consider an infinite haploid population, polymorphic at two loci, in the environment illustrated in Figure 9. It is supposed that selection acts on the haploid individuals, permitting *AB*, *Ab* or *ab* genotypes to survive according to which region they are in. Individuals may then migrate, as shown in the figure; they then fuse in pairs, and immediately produce new haploid offspring, with or without recombination, which start the cycle again.

There is no selection for or against recombination in the terminal regions, because double heterozygotes are not formed. However, in the central region *AB/ab* heterozygotes arise, and in these recombination is selectively advantageous, since it can produce *Ab* offspring of high fitness. Thus, in this rather artificial model, there is short-term selection for higher recombination in the central region and no selection elsewhere. The essential feature of the model which gives

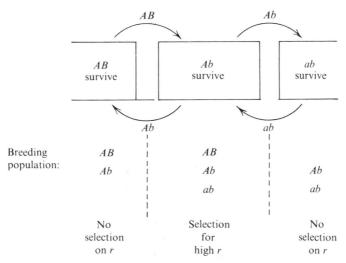

Figure 9. Migration in a spatially heterogeneous environment.

this result is that *AB* and *ab* individuals migrate far enough to meet in the central region, but not far enough to meet in one of the terminal regions. It therefore depends on the ratio between the distance an individual migrates and the size of the central region.

Now consider the more realistic case of a linearly arranged habitat with two sharp discontinuities a distance *Z* apart, at one of which selection switches from favouring *B* to *b*, and at the other from *A* to *a* (Figure 10). The typical individual migration distance is *l*. Then if

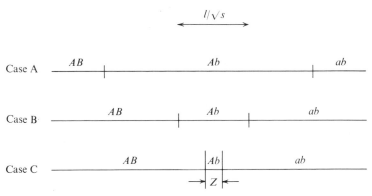

Figure 10. Relation between the typical characteristic migration distance $l/\sqrt{s}$, and the distance between environmental discontinuities *Z*. The typical distance moved by an individual is *l* and *s* is the selective intensity.

$Z \gg l/\sqrt{s}$, (case A), there will be two independent clines, no linkage disequilibrium and no selection on recombintion modifiers. If $Z \ll l/\sqrt{s}$ (case C), there will be a small central region in which selection favours higher recombination and much larger areas on either side of the discontinuities in which selection favours lower recombination. Finally, if *Z* and *l* are approximately equal (case B), the model corresponds roughly to that in Figure 9, and the main effect will be selection for increased recombination in the central region.

Slatkin (1975) analysed a continuous model of this kind, for a diploid model, with various values of the fitnesses and the recombination fraction. He showed that linkage disequilibrium will exist if the selective advantages are of the same order of magnitude or greater than the recombination fraction. He also suggested that there would be selection for higher and lower recombination as outlined above, although he did not prove it. There seems to be little doubt that the suggestion is correct, although it would probably require epistatic fitness interactions (the assumption of Figure 9, that only the optimal genotype survives, implies epistasis).

Spatial variation in the environment can select for increased recombination, just as temporal variation can. The assumptions which must be made are perhaps more plausible in the spatial case. However, it is equally easy to think of patterns of spatial variation which will select for decreased recombination. In particular, if the selective advantages switch over at the same point for two separate loci, selection will tend to reduce recombination between them.

C   Sib competition

The model described in Chapter 6, section A, was developed after discussions with Dr G. C. Williams in which he suggested that a combination of intense selection and an unpredictable environment would give a short-term advantage to sex. It led me to the conclusion that a special and rather implausible kind of unpredictability was required to produce the desired result. When his ideas were later published (Williams & Mitton, 1973; Williams, 1975), I realised that he had incorporated a feature which I had omitted; he supposed that the offspring of a single parent were in competition with one another.

The reason why this makes a difference is that asexually produced offspring resemble one another more closely, and therefore compete more severely, than do sexually produced offspring. The first formulation of the idea known to me is by Baker (1959), who wrote 'If a parent tree had formed a high proportion of its seeds by outcrossing, it is likely that there will be genetic differences between the competing offspring with an opportunity for the selection of that one which is best adapted to the prevailing conditions.'

Williams (1975) makes essentially the same point in a metaphor which is so vivid that it will bear repeating. An asexual parent, he says, is like a man who buys 100 tickets in a raffle with only one prize, and finds that they all have the same number. A sexual parent may be able to afford fewer tickets, because of the cost of producing males, but at least they all have different numbers. Note that there is only one winning number. That is to say, all the offspring of a single parent are competing in the same environment. If offspring were randomly distributed between environments (as was assumed in my 1971*a* model), then an asexual parent would be like a man who bought one ticket in each of 100 different raffles, all with different winning numbers. In this case he would not mind if all his tickets had the same number.

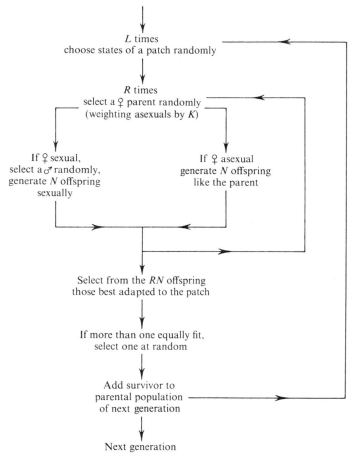

Figure 11. Flow diagram of sib-competition model. For symbols, see text.

Williams's 'elm–oyster' and 'aphid–rotifer' models have the
phenomenon of sib competition in common. Although they are
suggestive, I did not find them wholly convincing, because they are
not fully worked out. That is to say, it is impossible to follow one of
these models through a number of generations (or even through one
complete generation) to see how the frequency of different types
actually changes. For this reason I have developed a model of sib
competition (Maynard Smith, 1976*f*) which is intended to incorpor-
ate the essential features of Williams's models, and have investigated
its behaviour.

Sib competition is hard to treat analytically. The model has
therefore been simulated by 'Monte Carlo' methods; that is, by
simulation in which each individual is separately represented in the
computer. This is expensive in computer time, but has the compensat-
ing advantage of realism.

The model can be thought of as referring to a population of annual
plants of two types: sexual (hermaphroditic) and asexual. The
environment consists of L 'patches', each capable of supporting a

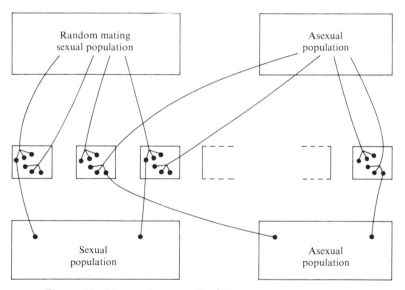

Figure 12. Dispersal pattern for sib-competition model, with $N = 3$
offspring per parent, and $R = 2$ parents contributing to each patch.

single adult. Each patch has five features, each of which can be in two states, $\bar{A}$ or $\bar{a}$, $\bar{B}$ or $\bar{b}$, . . . $\bar{E}$ or $\bar{e}$. In each generation the states of each patch are randomly assigned, the thirty-two possible types of patch being equally common.

Sexual individuals are diploid with five unlinked loci and two alleles at each locus, A or a, B or b, and so on. Genotypes AA and Aa are equally adapted to $\bar{A}$ and aa to $\bar{a}$; the remaining loci are similarly related to the other features. Thus an individual can be adapted to 0, 1, . . . 5 features of the environment; its fitness depends only on this number and on what competitors are present.

Asexual individuals produce offspring identical to themselves; they may have any one of the thirty-two possible genotypes, adapted to each of the possible types of patch.

The method of simulating the model is illustrated in the flow diagram (Figure 11) and in Figure 12. The essential points are as follows:

(i) Each patch receives N seeds from each of R parents.

(ii) The parents are chosen randomly from the L surviving adults in the previous generation, weighting the asexual parents by K to allow for the advantage of not producing pollen.

(iii) When RN seeds have been produced in a patch, one survivor is chosen randomly from those best adapted to the patch.

In biological terms, this procedure means that mating among sexuals is random and dispersion is random, except that seeds fall in 'packets' of N. A hermaphroditic breeding system was chosen for convenience of simulation; a dioecious system with $K=1$ would behave similarly to the present model with $K=2$ (the sexual females would suffer their twofold disadvantage through producing males, instead of through producing fewer seeds).

The method of selecting the survivor in each patch needs some justification, since it seems to imply that all deaths are genetic. An alternative interpretation is to suppose that each of R females disperses mN seeds to each patch ($m > 1$), and that these are reduced to N seeds by non-selective mortality before selection starts. RN then gives an upper limit for the total intensity of selection, $W_{max}/\bar{W}$; the actual value is usually lower because in most patches there will be more than one equally well-adapted individual from which a survivor is selected randomly. R measures the intensity of between-family selection and N of within-family selection.

Some results of simulating this model are shown in Table 3. In each case the population was started off with equal numbers of sexual and asexual individuals. Among the asexuals, the thirty-two possible clones were equally frequent. Among the sexuals, the frequency of the recessive alleles at each locus was initially 0.7, giving approximately equal frequencies of the thirty-two possible phenotypes. Each simulation was continued for ten generations; the final frequencies of sexuals are given in the table. The following conclusions can be drawn:

(i) If $N = 1$ (no sib competition), sex confers no advantage. This can be regarded as a confirmation of the view that mere environmental unpredictability is insufficient to confer an advantage on sex.

(ii) If $R = 1$ (only one family per patch), sex confers no advantage. Clearly, if one individual is certain to survive from each family, the genetic system is irrelevant.

(iii) If R and N are both greater than one, sex confers a selective advantage. If the final frequency of sexuals is greater than 0.5, the advantage is sufficient to outweigh the corresponding value of K. If sex is to outweigh the full twofold advantage of parthenogenesis, the critical value of RN is in the range thirty to forty; that is, each patch must receive thirty to forty times as many seeds as there is room for. It is important to remember that this quantitative prediction depends, among other things, on the assumption of five independent environmental features. This is one reason for desiring an analytical handle on the problem. However, it seems likely that Williams was correct in thinking that a very high total intensity of selection (one quite incompatible with the breeding system of most higher vertebrates) is required to overcome the twofold disadvantage of sex.

In each case, simulations were started with all thirty-two clones represented. This may be an unreasonable assumption. In all those simulations in which a substantial number of sexual individuals were present after ten generations, all five loci were still polymorphic, with frequencies of approximately 0.7 of the recessive alleles. However,

Table 3. *Results of simulating a model of sib competition. Proportion of sexual individuals after ten generations (initial value, 0.5) (from Maynard Smith, 1976f)*

Population size L	Intensity of selection			Advantage of asexual reproduction (K)					
	between families R	within families N	total RN	1.0	1.2	1.4	1.6	1.8	2.0
400	20	1	20	0.48	—	—	—	—	—
400	1	20	20	0.51	—	—	—	—	—
200	2	4	8	0.97	0.69	0.29	0.09	0.10	0
200	4	4	16	1.0	0.92	0.79	0.37	0.38	0.14
200	6	6	36	—	—	0.92	0.95	0.56	0.54
400	6	8	48	—	—	—	—	0.78	0.64

even in populations containing a substantial number of asexual individuals, not all clones were maintained. For example, in the simulation with $L=400$, $R=6$ and $N=8$, which started with 200 asexual individuals and ended with 146, only 20 of the 32 clones were present after 10 generations. This is the effect of random sampling in a finite population. In an infinite population, frequency-dependent selection would maintain all clones, just as it maintains polymorphism at all loci. But in a finite population clones are likely to be lost by genetic drift.

It is not necessary for all thirty-two clones to be present for the sexual population to be eliminated, but there must be a substantial proportion. When a single clone was introduced into a sexual population, it increased until it formed about 15% of the total population, but no further. This is a specific example of the general point made in Chapter 4, section F. Even with a twofold advantage, one clone is unlikely to drive its sexual ancestors to extinction over the whole ecological range.

The model can easily be modified to simulate selection for alleles for high and low recombination in a wholly sexual population. The assumption about the environment, and about dispersion, mating and selection are unaltered. Only diploid sexual hermaphrodites exist. There are now six linked loci. The first five, $AB \ldots E$ adapt individuals to features of the environment as before. The sixth locus has two alleles, R_L and R_H; the former is dominant for low recombination, and the latter recessive for high recombination. The loci arc linearly arranged in the sequence $ABCDER$. The recombination fraction between all pairs of neighbouring loci is C_H in R_H/R_H homozygotes, and C_L in R_H/R_L and R_L/R_L genotypes; interference is absent.

The results of simulating the model are shown in Table 4. The final column gives estimates of the selective advantage per generation of individuals homozygous for the high recombination allele.

The replicate runs in the table show that, for such a small population size, estimates of selection are only approximate. However, it is clear that there can be a substantial advantage for higher recombination. This advantage does not disappear if the recombination locus is unlinked to the fitness alleles. Also, in contrast to the

Table 4. *Selective advantage of recombination (Maynard Smith, 1976f)*

Population size L	Parents per patch R	Offspring per parent N	Total selection intensity NR	Recombination fractions C_L	C_H	Number of generations	Frequency of allele R_H Initial	Final	Estimated selective advantages of R_H/R_H
200	4	8	32	0.0	0.5	15	0.265	0.462	+0.165
200	5	10	50	0.05	0.5	10	0.262	0.387	+0.181
400	5	10	50	0.05	0.5	6	0.330	0.426	+0.182
400	5	10	50	0.05	0.25	6	0.301	0.388	+0.189
400	5	10	50	0.05	0.25	6	0.305	0.355	+0.115
Locus of R unlinked to fitness loci:									
400	5	10	50	0.05	0.5	6	0.316	0.375	0.127
400	5	10	50	0.05	0.5	6	0.272	0.325	0.147

models to be discussed in Chapter 7, there is still substantial selection for higher recombination even when there is already 5% recombination between neighbours.

So far, then, the model strongly supports Williams's argument that with sib competition there can be selection for sex and for higher recombination. Unfortunately, however, there are two serious difficulties.

Many loci are concerned with a single environmental feature

Suppose that alleles at two loci are concerned with adaptation to a single environmental feature; for example, alleles A_1 and A_2, at different loci, adapt the individual to high temperatures, and a_1 and a_2 to low temperatures. Then genotypes of high fitness will be either A_1A_2 or a_1a_2, and recombination will be disadvantageous.

Exactly the same result arises if we suppose that only one allele is concerned with each environmental feature, but that there is an association between states $(d \neq 0)$. Thus, if hot places are always dry and cold places wet, this is formally equivalent to there being only one feature with two states, 'hot and dry' or 'cold and wet', with two loci concerned with one feature.

As Figure 13 shows, the effect can be an important one. If, instead

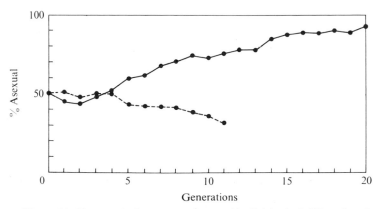

Figure 13. Changes in frequence of asexual individuals: full line, fourth and fifth environmental features identical; dashed line, all five environmental features independent. $L = 400$, $R = 6$, $N = 8$, $K = 2$.

of five independent features, there are four, and two loci concerned with one of them, this converts a small advantage for sexual reproduction into an advantage for asexual reproduction.

The objection is serious, since it must often be the case that alleles at several loci are concerned with adaptation to a single environmental feature. Selection will tend to reduce the rate of recombination between loci concerned with the same feature, and (if there is sib competition) increase it between loci concerned with different features. These tendencies are incompatible unless the sequence of loci, and their association in linkage groups, can be changed. In reality, there seems little evidence for association between loci with related functions, except for the special case of 'supergenes' discussed in Chapter 5, section D. The rarity of such associations is presumably because chromosomal rearrangements occur rather infrequently, and often cannot easily be incorporated by selection because of their effects on fertility.

Patterns of dispersion

If sib competition is to favour higher recombination, the dispersal patterns of the species must have two features; offspring must not find themselves in an environment similar to that of their parents, yet they must compete with one another. This seems to rule out species such as the oyster, in which dispersal is so wide that it is unlikely that two sibs will settle close enough to one another to compete. It may also rule out most plant species, since most seeds can fall near the parent plant. This latter objection disappears if the environment in one place changes drastically between one generation and the next, as may well be the case, particularly if the mere presence of an individual of a particular type renders the environment less suitable for another individual of the same type.

D   Environmental unpredictability – the evidence from geographical parthenogenesis

If sexual reproduction and a high rate of recombination have evolved as adaptations to an unpredictable environment, this should be reflected in the geographical distribution of parthenogenesis. Levin (1975), basing himself in part on earlier reviews by Stebbins (1950), Grant (1958) and Baker (1959), makes the following generalisations about the breeding system of plants:

(i) Weedy and colonising species are often apomictic, or self-fertilising hermaphrodites.

(ii) Recombination-restricting devices (low chromosome number, localised chiasmata, obligate heterozygosity) are commonest in plants of transient, temperate habitats.

(iii) Within species, the frequency of self-compatibility and self-fertilisation is higher in marginal than in central populations.

In contrast, parthenogenesis and self-fertilisation are rare in tropical environments. Levin explains these observations as follows. There is no point in a weedy species attempting to track changes in the physical environment by genetic change, precisely because these changes are unpredictable. A population which adapts genetically to its current habitat is likely to be less well adapted after dispersal to a new habitat. In contrast, the physically uniform environment of the tropics is rich in pathogens and herbivores, which are themselves continuously evolving new capacities to attack plants. A plant species which could not evolve new resistance mechanisms could not survive. A weed can escape its enemies by dispersal; a plant in a more stable community must evolve to meet ever-changing threats.

An essentially similar empirical generalisation, and a similar explanation, had been proposed for animals by Glesener & Tilman (1978). Following up an earlier suggestion by Suomalainen (1950), they show, from a review of terrestrial animals, that parthenogenetic races, when compared to their nearest sexual relatives, tend to occur

(i) at higher altitudes and latitudes,
(ii) in xeric as opposed to mesic conditions,
(iii) in disturbed habitats, or
(iv) on islands or in island-like habitats.

These ecological correlations are in part explained by the fact that parthenogenesis (and self-fertility in hermaphrodites) is an advantage in a colonising species, since a single individual can establish a colony.

However, both Levin and Glesener & Tilman argue that there is more to the correlation than this. They suggest that the one feature which which these situations have in common is that there are relatively few competing, predatory, or pathogenic species. The authors suggest that sex and genetic recombination are necessary adaptations of any species which lives in a biologically complex environment. Other species in the ecosystem will evolve and, therefore, any species which is to survive must itself evolve to meet the challenge. The similarity between this and Levin's argument based on plant data is obvious. Both, in effect, accept Van Valen's Red Queen hypothesis (see p. 23); the environment of each species deteriorates, not primarily because of physical changes, but because other species in the ecosystem evolve.

I find it persuasive that a comparative study of plants and of animals should have led independently to the conclusion that sex and recombination are most essential in complex ecosystems in physically uniform environments. The empirical fact that parthenogenesis is commonest in colonising species is fully in accord with the theoretical argument that an unpredictable environment does not in itself favour sex or recombination. What is characteristic of a biologically complex and changing environment is the continuous appearance of qualitatively new selective forces. It is not simply a matter of ringing the changes within a constant set of possibilities, as in the models in this chapter. If that were all, we would be hard put to it to account for sex, except in those cases in which sib competition is an appropriate model. But what of an environment in which qualitatively new challenges are forever arising? This is the subject of the next chapter.

7 Short-term advantages for sex and recombination – II. Selection in a finite population

Preamble

In this chapter, I explain what I believe to be the most significant process maintaining sex and recombination in nature. Two model systems are described in which there is selection for higher recombination. At first sight the models may not seem to have much else in common, but a more careful examination shows that they depend on the same basic mechanism. What is important, therefore, is not the details of the model but the underlying mechanism. As it happens, the authors of the two models, Strobeck *et al.* (1976) and Felsenstein & Yokoyama (1976) visualise this mechanism in rather different ways. The first model was conceived as an illustration of the 'hitch-hiking' phenomenon (Maynard Smith & Haigh, 1974), and the second as an illustration of the 'Hill–Robertson' effect (see p. 32). This is not so much a difference of opinion as a different way of putting a mathematical argument into words.

Methods of verbalising are important. To be able to present a formally correct mathematical model in which alleles for higher recombination increase in frequency is only half the battle. It is then necessary to get an intuitive understanding of why the increase happens, because only then can one have an idea of the range of circumstances in which the mechanism is likely to operate. This is particularly important when, as in the case of both models, analysis has been by computer simulation. Accordingly, in section A I shall first digress to explain the concept of hitch-hiking in as general terms as possible. Section B then describes the model of Strobeck *et al.*, interpreting the results in terms of hitch-hiking, and section C does the same for the model of Felsenstein & Yokoyama. In section D, I digress to discuss the effects of selfing on the intensity of selection for recombination in the models discussed in this chapter and the last.

The reason for this digression is that there is empirical evidence that the rate of recombination is higher in naturally selfing populations, and it is, therefore, interesting to ask whether the various models can predict this effect. Finally, in section E, the models of this chapter are looked at from a different point of view, in which recombination is seen as being favoured because of randomly arising linkage disequilibrium, and an attempt is made to reach some general conclusions.

A   Gene selection and hitch-hiking – a digression

The clearest introduction to the phenomenon of hitch-hiking is through Cox & Gibson's (1974) experiments on *Escherichia coli*. Two non-recombinant strains of bacteria, which differed only in that one of them carried a 'mutator' gene greatly increasing the rate of mutation over the wild-type, were allowed to compete. The high mutation rate strain was started at a low frequency, but in every case increased in frequency until it replaced the wild-type strain. Why should this be so? Since most new mutations are likely to be harmful, one might expect the mutator gene to decrease in frequency. The authors were able to show that the explanation is as follows. In the abnormal environment of the chemostat, there must be many mutations which, if they occur, would increase the fitness of their carriers. Such favourable mutations are likely to occur first in the mutator strain. Once such a mutation has occurred in the mutator strain, that strain will increase in frequency at the expense of the wild-type strain and will ultimately replace it. Since there is no genetic recombination between the strains, an increase in frequency of the new favourable mutation causes a parallel increase in the mutator gene which caused the mutation.

This can be expressed by saying that the mutator gene obtains a hitch from the new mutation; some of the effects of hitch-hiking in diploid populations have been analysed by Maynard Smith & Haigh (1974) and Thomson (1977). The concept seems particularly useful when thinking about 'second-order selection' on genes altering mutation or recombination rates. Thus, such genes may have no effect on the phenotype of the individual. We should not, therefore, analyse their changes in frequency by asking what effect they have on the fitness of individuals, since they have no such effect. The concepts of individual and group selection are in fact not particularly helpful in this context. Instead, we should remember that it is genes and not individuals which are replicated and transmitted. Consequently, we should ask questions, not about individual survival, but about the number of copies of a particular gene which will be present at some time in the future.

Williams (1966) argued that we should think in terms of the selection of genes, and Dawkins (1976) has suggested that this approach is required if we are to understand the evolution of sex. The main suggestion which I am making in this chapter is as follows. Genes for high recombination or mutation alter the sets of alleles at other loci with which they will be associated in future generations. If they alter these sets in such a way that, on average, they associate with genes or combinations of genes conferring high fitness, then they themselves will increase in frequency.

Returning for a moment to Cox & Gibson's experiment on *E. coli,* it is worth asking why the hitch-hiking effect of new favourable mutations in an upward direction outweighed the effects of the far more numerous harmful mutations in a downward direction. The answer is as follows. Suppose that a population of 10^9 bacteria contains 1000 individuals with the mutator gene. A favourable mutation occurs in one of these individuals, and unfavourable mutations in the other 999. The latter leave no descendants, but descendants of the former increase in frequency until all 10^9 bacteria carry the favourable mutant and, because there is no recombination, also carry the mutator gene to fixation. The effect depends on the absence of recombination; having hitched a ride, the mutator gene does not let go.

There is, of course, a good reason why geneticists have preferred to think in terms of the fitness of genotypes and to calculate gene frequency changes in terms of those fitnesses, rather than to think directly in terms of the genes themselves as the units of selection. What we observe are the phenotypes of individuals, which are the product of interaction between genotype and environment. We would like to understand the evolution of phenotypes, in terms of the ways in which a particular structure or behaviour influences survival.

It is, therefore, natural to ascribe fitnesses to particular genotypes in particular environments. Hamilton's (1964) concept of 'inclusive fitness' can be seen as a way of preserving this method of approach, while allowing for the fact that an individual's behaviour may affect not only its own survival, but that of other individuals carrying identical genes.

There does not, however, seem to be any good reason for sticking to this method of thinking when the genes themselves do not (or at least need not) affect the fitness of the individual in which they find themselves, but only alter the genes with which they will be associated in future generations. The model discussed in the next section works because a gene causing higher recombination first produces a new gene combination of high marginal fitness, and then hitches a ride.

B   Hitch-hiking and recombination

Consider the following situation. A diploid population has a balanced polymorphism for alleles A,a. At a closely linked locus a favourable mutation B occurs. An increase in frequency of B is interfered with by selection at the A locus. Will selection favour an allele C^+ at a third locus, causing recombination between A and B, and in this way reducing the interference?

It will be convenient, first, to consider a particular numerical example. Two alleles are segregating at each of three loci in a random-mating diploid population, as follows:

(i) A pair of alleles A,a are maintained in a balanced polymorphism; the fitness of AA, Aa and aa are 0.5, 1.0 and 0.5, respectively.

(ii) A favourable, partially dominant allele B is replacing its recessive allele b; the fitnesses of BB, Bb, and bb are 1.2, 1.0, and 0.8, respectively. Fitnesses between loci are multiplicative.

(iii) A pair of alleles C^+,C^- affect recombination between the other loci; C^+ is a recessive allele for recombination. The recombination fraction between A and B is 0.01 in C^+/C^+ homozygotes, and zero in C^+/C^- and C^-/C^-. In all three genotypes there is 0.01 recombination between B and C.

Initially, A and a have their equilibrium frequency of 0.5; C^+ has an initial frequency of 0.2, and there is linkage equilibrium between the loci of A and C. B has an initial frequency of $1/2000$, and occurs only in AB gametes; this corresponds to a single B mutation, which happens to occur on an A chromosome, in a total population of 1000 individuals.

The subsequent changes in frequency of the genes were then calculated deterministically, and are shown in Figure 14. Despite the deterministic calculation of gene frequency change, the model essentially refers to a finite population because it assumes that a single unique mutation to B took place; in an infinite population, AB and aB gametes would both arise by mutation, in equal frequencies.

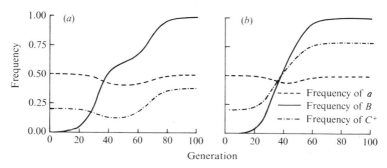

Figure 14. The hitch-hiking effect on a gene for recombination: *(a)* Gene frequencies when B is initially in coupling with C^-; *(b)* gene frequencies when B is initially in coupling with C^+. Frequencies of a, – – –; of B, ——; of C^+ — · —. (From Strobeck *et al.*, 1976.)

Consider first changes at the A and B loci. The frequency of B increases for a time, but then stabilises, because, in the absence of aB chromosomes, selection at the A locus prevents further frequency change. In time, aB chromosomes arise by recombination, and B resumes its increase to fixation. The frequency of A first increases, and then, when aB chromosomes are present, falls back to its equilibrium frequency of 0.5.

At the C locus there are two cases to consider. If the original B mutation occurs in coupling with C^-, giving an ABC^- chromosome, the allele C^+ initially falls in frequency (Figure 14a); if B occurs in coupling with C^+, giving an ABC^+ chromosome, the frequency of

C^+ initially rises (Figures 14b). In both cases, since recombination between A and B occurs only in C^+/C^+ homozygotes, the aB gametes arise by recombination from ABC^+/abC^+ individuals, giving an aBC^+ gamete. As the newly arisen aB chromosome increases in frequency, it carries the C^+ gene up in frequency as well. Thus, during the fixation of the favourable mutant B, the allele C^+ experiences two hitch-hiking events: the first may increase or decrease its frequency, according to the chances of whether B occurs first in coupling or in repulsion; the second hitch-hiking event always increases the frequency of C^+.

We calculated the expected change in frequency of C^+ as follows. With probability 0.2, B occurs in coupling with C^+, and the final frequency of C^+ is 0.787; with probability 0.8 B occurs in repulsion, and the final frequency is 0.383. Hence the expected frequency of C^+ is $0.2 \times 0.787 + 0.8 \times 0.383 = 0.464$; the initial value was 0.2.

This numerical example has been described in some detail, because it illustrates the essential process. An allele for high recombination produces a selectively favourable chromosome, aBC^+, and thus hitches a ride to a higher frequency. Strobeck *et al.* (1976) report the results of simulations with varying parameter values, as follows. If Δp is the expected change in the frequency of C^+, the allele for high recombination, then:

(i) Δp is zero or positive.

(ii) Δp varies in a complex way with the selective forces at the A and B loci; it is not necessary that the force be greater at the A than at the B locus, as in the example.

(iii) The effect occurs whatever the mechanism maintaining the polymorphism at the A locus; for example, it could be maintained by frequency-dependent selection.

(iv) Δp is small unless there is reasonably close linkage between the three loci in the absence of allele C^+. This is particularly so for linkage between A and B; Δp is negligibly small unless the recombination fraction in C^-/C^- is less than 0.01. Linkage between B and C is less critical; Δp may still be appreciable when recombination between B and C is 0.1, but is near zero for free recombination between B and C.

(v) The largest values of Δp occur when C^+ is recessive, and when recombination between A and B is low in the presence of C^-. However, Δp may still be substantial when C^+ is dominant.

Thus, if selective substitutions are occurring at some loci, while at others balanced polymorphisms are maintained, there will be selection for increased recombination. The main factor which limits the importance of this process is (iv) above. If the effect is to be substantial, then in the absence of high recombination alleles there must be close linkage between the selected loci, and between them and the loci influencing recombination. This is because C^+ alleles only get a substantial hitch if they stay linked to the favourable gametes they produce. Thus the model answers Turner's (1967) question 'why doesn't the genome congeal?', but does not explain why the genome is as loose as it is.

B   Selection for recombination in the presence of recurrent mutations

Felsenstein & Yokoyama (1976) start their paper by making an obvious but illuminating point. Fisher (1930) and Muller (1932, 1964) showed that recombination would be favoured by inter-population selection. There are special circumstances in which their arguments can be used to show that recombination would be favoured by intra-population selection. Two conditions must be satisfied:

(i) The survival or extinction of populations must depend on the fitnesses of the individuals composing them.

(ii) If we divide the haploid genomes, or gametes, of the population into two sub-populations according to whether they do or do not carry the allele C^+ for high recombination, then there must be no gene flow between the two sub-populations.

The second of these conditions is satisfied if C^+ is recessive, and if there is no recombination in C^-/C^- homozygotes. Then, by the Fisher–Muller argument, the C^+ sub-population will accumulate favourable mutations and eliminate unfavourable ones more rapidly, and will selectively eliminate the C^- sub-population. In hitch-hiking terms, the C^+ allele produces favourable gene combinations, and is carried to fixation by them.

It is less obvious what will happen if the C^+ allele is dominant, because in this case it will let go of the favourable gene complexes it has brought into being. It is also not clear what will happen if there is some recombination in C^-/C^- homozygotes. Felsenstein & Yokoyama attempted to answer the first of these two questions by Monte Carlo simulations of small populations.

In the first series of simulations, they investigated the effects of recurrent favourable mutations. They considered haploid populations of size $N = 100$ and 200, half the individuals initially carrying the C^+ allele and half C^-. Favourable mutations occurred at a rate u per genome per generation, each mutation occurring at a new locus. The relative fitness of an individual with k favourable mutations was taken as $(1+s)^k$. Each generation was produced as follows. Two parents were selected randomly from the previous generation. If both were C^+ (or, in simulations in which C^+ was dominant, if one parent was C^+) a single offspring was produced by free recombination between all loci; otherwise a single offspring was produced which was identical to the first parent. This procedure was repeated N times.

Each simulation was continued until either C^+ or C^- was fixed. Selection favouring C^+ was demonstrated if C^+ was fixed in significantly more than 50% of the cases.

A second series of simulations investigated recurrent deleterious mutations, with $N = 50$ and 100; the simulation was essentially similar in other respects.

For favourable mutants, C^+ was fixed more often than C^-, both when recessive and when dominant. The former result could have been predicted with some confidence, since no gene flow occurred between C^+ and C^- populations. It was less obvious that C^+ would increase when dominant. Simulations were run with u from 0.1 to 1.0 for unfavourable mutations, and from 0.001 to 0.01 for favourable ones, and $s = 0.25, 0.5, 1$ and 3. Selection for C^+ was greatest for large u, and, when C^+ was dominant, for large s.

For deleterious mutants, C^+ was fixed more often than C^- when it was recessive, but there was no significant difference when C^+ was dominant. This result could have been predicted provided it could be shown that Muller's ratchet was operating in the C^- sub-population. It was shown in Chapter 3 (p. 34) that the ratchet will operate if $n_0 = Ne^{-u/s}$ is small. This condition was certainly satisfied in the simulations; in fact, significant selection in favour of C^+ was observed only when $u/s \geqslant 2$.

Thus, the simulations with deleterious mutations do not add much to what could be predicted analytically. With favourable mutations, however, it was shown that even a dominant allele giving free recombination is selected for, in place of a recessive allele giving zero recombination. This is a little difficult to see in hitch-hiking terms. Thus suppose at some moment the population contains $m_1 +$, $+ m_2$, and $+ +$ gametes, where m_1 and m_2 are favourable alleles. An $m_1 m_2$ gamete can arise by recombination from either

$$\frac{m_1 + C^+}{+ m_2 C^+} \quad \text{or} \quad \frac{m_1 + C^+}{+ m_2 C^-},$$

or other genotypes equivalent to these. In the former case the new gamete will be $m_1 m_2 C^+$, and C^+ will receive a hitch, but only a short one; in the latter case the new gamete could be $m_1 m_2 C^-$. It would be very desirable to have a better intuitive grasp of why, in these simulations, C^+ was fixed 1787 times, and lost only 813 times.

It hardly needs to be added that this idea was first adumbrated by Fisher (1930, p. 118). Discussing what factors might oppose selection for tighter linkage, he suggests that if several favourable mutants are spreading simultaneously through a population, a gene for recombination could benefit by producing chromosomes carrying two favourable mutations. He adds 'It is apparent that, for this process to have been an effective check upon the constant tendency to increase the intensity of linkage, the stream of favourable mutants must be an abundant one.'

D   Effects of selfing on selection for higher recombination

One of the empirical generalisations about recombination which appears to be reasonably well established is that the frequency of recombination tends to be higher in selfing or inbreeding species or varieties than in related outbreeding ones (for references, and a summary of the evidence, see Charlesworth, Charlesworth & Strobeck, 1977, and Lewis & John, 1963). This is probably not a mere physiological consequence of genetic homozygosity; indeed, Rees (1956) showed that in rye the immediate physiological consequence of inbreeding is to decrease the chiasma frequency and to increase its variance. The difference between wild varieties, therefore, can be taken as being a consequence of selection acting on the genetic consequences of recombination.

It is, therefore, worth asking whether the various models discussed in this and the previous chapter predict that selection for higher recombination will be more intense in partially selfing populations. This question was investigated by Charlesworth *et al.* (1977). Before describing their results in detail, I want to warn against an oversimplistic approach to the problem.

This approach is to argue as follows. In a selfing population the opportunities for recombination between alleles are reduced, because most individuals are homozygotes. Selfing has the same effect in reducing the rate of approach to linkage equilibrium as does linkage. Hence, if there is an optimal rate of recombination between two loci in an outcrossing population, the optimal rate in a selfing population will be higher, to make up for the reduced number of opportunities.

Two things are wrong with this argument. First, it fails to distinguish between situations in which the equilibrium recombination rate is altered by selfing and those in which it is the same. Thus if the equilibrium recombination rate is higher with selfing, then selection for increased recombination may also be greater, and if it is lower the reverse will be the case. But in many cases the equilibrium recombination rate may be unaltered by selfing – this is often so when the equilibrium rate is zero or 50%. In these cases, there may still be a different intensity of selection with selfing, but it is hard to predict whether selection will be more or less intense. A second weakness, of course, is that the argument is implicitly group-selectionist, in that the optimum rate may be thought of as an optimum for the population.

The results obtained by Charlesworth *et al.* were as follows:

(i) *The hitch-hiking model of Strobeck* et al. *(1976)* in which the progress to fixation of an advantageous mutant is inhibited by linkage to a polymorphic locus. If the selected loci *A* and *B* and the recombination locus *C* are all closely linked, selfing has little effect. But if *C* assorts independently of the selected loci, the increase in frequency of the high recombination allele is vanishingly small in the absence of selfing, but can be substantial with 80% selfing. Also, if *A* and *B* recombine with a frequency of 1% or more, selection in the absence of selfing has little effect, whereas with 80% selfing there is still a substantial effect when the *A–B* distance is 10%. Intuitively, the importance of selfing in this model is that the recombination allele which produces a favourable gamete can hang on to it for longer.

(ii) *The model of Felsenstein & Yokoyama (1976)* in which different favourable alleles interfere with one another. If in the absence of a recombination modifier there is *no* recombination between two favourable mutants *A* and *B,* then, as Felsenstein & Yokoyama found, there is always some selection for a modifier increasing recombination. But if there is already some recombination between *A* and *B,* selection may actually reduce it. In most cases, if there is selection for increased recombination, its extent is increased by selfing.

(iii) *The model of Charlesworth (1976)* of selection in a fluctuating environment. In this model, there is an equilibrium value of recombination frequency, which may be 50% or lower. In general, the equilibrium value is higher in the presence of selfing. Selection for alleles increasing the frequency of recombination is greatly increased in the presence of selfing.

(iv) *Selection in a uniform environment.* In this case, of course, selection tends to reduce recombination frequency. The intensity of this selection may be increased by selfing. The reason for this can be seen by considering the extreme case in which homozygotes for either allele at either locus are lethal. Then an *AB/ab* individual which selfs will produce 50% viable offspring if there is no recombination, but 25% if there is free recombination.

It is clear that the effects of selfing on selection for recombination are complex. Either the hitch-hiking or the fluctuating environment model could account for the fact that recombination is higher in selfing populations because, for both types, there is a wide range of parameter values for which selection for increased recombination is more intense in the presence of selfing. The authors did not analyse a sib-competition model from this point of view.

E   Conclusions

Felsenstein & Yokoyama (1976) suggest that there may be only two conceptually different models for the evolution of recombination. The first was discussed in Chapter 6, section A. Recombination will be favoured if the environment changes in such a way that the correlations between selectively relevant features of the environment change sign between generations. If this were so, selection would favour recombination even in an infinite population with random mating and dispersal. But it is hard to believe that the world is like that; to modify Einstein, God does not change the rules.

The second model rests on the fact that, in a finite population, chance events will generate linkage disequilibrium. As a result, alleles at different loci cannot respond to selection independently of one another. Frequency changes at one locus interfere with changes at another. One favourable allele cannot be fixed because of selection at another. Deleterious alleles accumulate as Muller's ratchet turns. Felsenstein (1974) referred to this as the 'Hill–Robertson effect'. His theme in that paper was that it is this randomly generated disequilibrium which confers an advantage on recombination. At that stage, Felsenstein was arguing in explicitly group-selectionist terms; he was talking about the ways in which the properties of populations are altered by recombination.

The purpose of the simulations described in section C was to see whether the advantages which were originally seen as group advantages could operate at the individual level in the short term. Clearly, the absence of recombination can lower the mean fitness of a population. However, an allele which would increase the mean fitness of a population does not necessarily increase in frequency. But mean fitness can give us a hint. Felsenstein's approach was as follows: let us first look at those situations in which an absence of recombination does lower the mean fitness. It turns out that the common feature of these situations is the random generation of linkage disequilibrium in finite populations, which had earlier been investigated by Hill & Robertson (1966). Let us then ask whether, in these situations, there is individual, short-term, selection for genes increasing recombination.

For whatever reason, it is clear that Strobeck *et al.* and Felsenstein & Yokoyama, have converged on a common model. In the former case, it was interference between directionally selected and stably polymorphic loci which favoured recombination; in the latter, it was interference between favourable genes, or between deleterious ones. Two questions remain. Are Felsenstein's two fundamental types of model the only possible ones? If so, are they sufficient to explain recombination?

It is too early to be confident of the answer to either question. Are there other models? Felsenstein & Yokoyama see the sib-competition models discussed in Chapter 6, section B, as yet another example of the Hill–Robertson effect. Certainly they depend on a finite number of parents contributing offspring to a single patch. But I cannot see that they are fundamentally similar. When one asks whether sib competition could explain the maintenance of recombination in any particular case, one is led to questions concerning patterns of dispersal and the relation between gene loci and environmental variation which are quite different from those which arise when thinking about the Hill–Robertson effect and hitch-hiking. For the present I prefer to see sib competition as a third type of model.

Are the models we have sufficient? I gave reasons at the end of the last chapter for doubting whether the environmental unpredictability and sib competition models are really sufficient, although both may play a role. What of hitch-hiking and random linkage disequilibrium? This model, although at present represented by two very special simulations, has the universality we seek. It must operate in all real populations at all times. But it has one grave weakness. It explains why some recombination (either a little, or 50%) should be favoured over no recombination. It does not as yet explain why a lot of recombination should be favoured over a little.

8 *Hermaphroditism, selfing and outcrossing*

Preamble

In this chapter I discuss the selective forces responsible for the evolution of hermaphroditism as opposed to dioecy, and of selfing and inbreeding as opposed to outcrossing. The topic is difficult because of the variety of different selective forces involved. In particular, we have to take into account:

(i) The allocation of resources to male and female functions.
(ii) The relative fitness of inbred and outbred offspring.
(iii) The probability of pollination, or of fertilisation.
(iv) The distribution, life history and ecological adaptations of the species concerned.

Sections A, B, and C are concerned primarily with models. Section A analyses the selection for self-incompatibility in hermaphrodites, because of the deleterious effects of selfing. One conclusion of the analysis is that we would expect to find phenotypic variability for the degree of selfing in populations, and the observational evidence bearing on this is presented. Section B analyses the patterns of resource allocation between male and female functions and the resulting selection for the hermaphroditic or the dioecious habit. Section C discusses three other models which have been proposed, and their relationship to the models already discussed. The main conclusion of this theoretical discussion is that there are three major factors which must be taken into account – the low fitness of inbred offspring, the advantages of self-fertility if a mate is not available, and the pattern of resource allocation.

In sections D and E, the distribution of hermaphroditism in plants and animals is briefly reviewed; this review confirms that all three factors are important, but leaves many questions unanswered.

Section F describes some of the mechanisms which have evolved in animals because of their role in preventing inbreeding, and section G the possible relevance of these mechanisms for our own species. It may be helpful to add a word on terminology. I have decided, reluctantly, to abide by zoological usage when speaking of animals and botanical usage when speaking of plants. Thus, animals with separate males and females are gonochoristic, whereas plants with this habit are dioecious. A hermaphrodite animal is one which produces both eggs and sperm; it may do so simultaneously or sequentially. A hermaphrodite plant is one with flowers which produce both ovules and pollen; a plant which produces separate male and female flowers on the same individual is monoecious. However, when discussing theories which apply equally to animals and plants I shall use hermaphrodite to mean an individual in which both sexes are combined, or a species composed of such individuals, and dioecious to mean a species in which the sexes are separate. I hope the context will make the meaning clear.

A   Selection for self-compatibility in hermaphrodites

If there were no differences between the fitnesses of inbred and outbred individuals, a hermaphrodite which fertilised itself would have an immediate advantage over one which did not. Thus, consider a population of self-incompatible hermaphrodites. On the average, if the population size is constant, each individual contributes two genomes to the next generation – one as a male and the other as a female. A rare mutant which was capable of selfing would, on average, contribute three genomes, two to its selfed offspring and one as a male by outcrossing. Yet some hermaphrodite animals (e.g. the snails *Helix* and *Cepaea*) cannot fertilise themselves, and plants have evolved a range of mechanisms (self-sterility alleles, heterostyly, protandry) which either prevent self-pollination or reduce its frequency. It is clear that there must be some short-term advantage to outcrossing and the obvious one is that offspring produced by outcrossing are usually fitter than those produced by selfing.

Imagine a hermaphrodite plant. How much less fit must selfed offspring be if self-incompatibility is to be maintained? Suppose that selfing is caused by a recessive mutant m, such that m/m individuals always self and $m/+$ and $+/+$ never do. Let the fitness of individuals produced by 1, 2, 3 . . . generations of selfing be V_1, V_2, V_3 . . ., compared to a fitness of 1 for offspring produced by outcrossing. These fitnesses can be thought of as either affecting probability of survival or causing a proportionate reduction in the number of seeds and pollen produced. Finally, let f be the proportion of ovules which are fertilised in a self-incompatible plant; it is supposed that in a self-compatible plant all ovules are fertilised. It can be shown (Maynard Smith, 1977b) that a rare mutant m will increase in frequency if

$$\alpha = 1 + \frac{V_1}{f} + \frac{V_1 V_2}{f^2} + \frac{V_1 V_2 V_3}{f^3} + \ldots > 2. \tag{8.1}$$

If the fitnesses of all the selfed generations are the same, and equal to V, this reduces to $(1 - (V/f))^{-1} > 2$, or $V/f > \frac{1}{2}$.

Condition (8.1) is for the initial increase of a rare gene for self-compatibility. It is hard to treat the subsequent increase analytically. It can, however, be shown that if there is a constant value of V, then for $V/f > \frac{1}{2}$, the self-compatibility gene will increase to fixation.

It is quite possible that $a > 2$, so that the self-compatibility allele will enter the population, and yet the allele will not increase to fixation. This will be so if fitness does not decline too seriously after one or two generations of selfing, but declines almost to zero if selfing is continued for more than a very few generations.

In effect, there are three possibilities:

(i) $a < 2$, as is in fact the case for the data given in Figure 15. Selection will favour complete self-incompatibility.

(ii) $a > 2$, but V_i (the fitness after i generations of inbreeding) very low for some i. Selection will favour either genetic polymorphism for selfing or a genotype which permits some proportion of selfing.

(iii) $a > 2$, and V_i not very low for any value of i. Selection will favour selfing.

So far I have discussed the evolution of selfing from outcrossing. What of the reverse transition? As before, let V be the fitness of a plant produced by selfing, compared to 1.0 for an outcrossed plant,

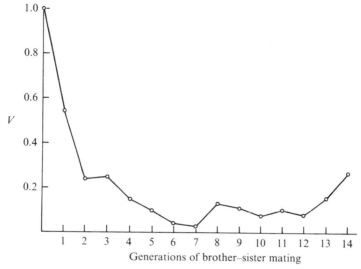

Figure 15. Average number (V) of surviving offspring per day per pair (relative to a value of 1.0 for the initial outbred population) in an inbred line of *Drosophila subobscura*. (Hollingsworth & Maynard Smith, 1955.)

and f be the chance of pollination of a self-sterile plant. A rare dominant mutant M arises in a population of selfing plants, such that $M/+$ plants do not self. The $M/+$ plants have f/V times as many offspring as selfing plants, and transmit the M gene to half of them. Hence the gene for self-sterility will spread if $V < f/2$.

The same condition, $V < f/2$, can be shown to hold for the spread of a self-sterility gene in a self-compatible species with a high frequency of outcrossing (achieved, e.g., by monoecy or protandry). In practice, self-sterility is more likely to arise in such a species than in a purely selfing one, for two reasons:

(i) A selfing species will produce little pollen, and so f will be small.

(ii) A selfing species will be better adapted to inbreeding, so that V will be large.

One conclusion of the preceding analysis is that we would expect to find that many populations have an intermediate frequency of selfing, either through genetic polymorphism for self-sterility or because of partial self-sterility in individuals. There is evidence that this is so. Allard, Kahler & Clegg (1977) have measured the frequency of selfing in plant populations by looking at isozyme frequencies in seedlings raised from seeds collected in nature. A deficiency of heterozygotes compared to the Hardy–Weinberg expectations indicates inbreeding, which is in the main a consequence of selfing. Their results for six species are shown in Table 5. *Lolium multiflorum* is a typical incompatible rye grass, yet in some populations over 30% of seeds are produced by selfing. *Avena barbata,* the wild oat, is a typical selfer, yet up to 7.5% of seeds may be outcrossed. *Collinsia sparsiflora* varies from complete selfing to almost random mating.

Table 5. *Outcrossing rates* (t) *determined from enzyme data in several plant species (from Allard* et al., *1977)*

Species	Range in *t* (%)	Number of populations
Avena barbata	0.1–7.5	21
A. fatua	0.1–1.6	9
Bromus mollis	7.0–14.0	2
Lolium multiflorum	68.0–104.3	11
Collinsia sparsiflora	0.0–84.0	11
Hordeum vulgare	0.0–8.5	3

Considering *Lolium* in a little more detail, Allard and his colleagues found that most plants will set some selfed seeds, and some plants are moderately self-compatible. Although more work needs to be done, it seems clear that at least some of the between-population differences shown in the table are genetic. Four populations from a transect 100 metres long showed different frequencies of outcrossing. The terminal populations, from flat, well-watered areas, had lower rates of selfing than the intermediate populations from a dry slope. The same association between high rates of selfing and xeric as opposed to mesic habitats was observed also in *Avena* and *Collinsia*; this agrees with the generalisation made on p. 109.

The most unexpected feature of the *Lolium* data is that estimates of rates of selfing for five different loci in a single population vary from

1% to 24%. This would be explained if there were marked linkage disequilibrium between the enzyme loci and the self-incompatibility locus, but only with some difficulty.

It is worth discussing a little further the likely genetical consequences of protracted inbreeding. Figure 15 shows that in *Drosophila subobscura* the mean fitness fell to a very low level after six to seven generations of brother–sister mating, but then rose again. In the next 200 generations it never again fell to the very low initial level, but neither did it rise to equal that of typical outbred flies. I believe this to be the usual picture of an inbred line in a species which outbreeds in nature (see, e.g., Lerner, 1954).

If the deleterious affects of inbreeding were wholly the result of lethal recessives, inbreeding would soon eliminate these lethals, and the fitness of the line would rise to equal that of a typical outbred population. There are two explanations for the fact that this does not happen:

(i) At some loci, both homozygotes are lower in fitness than the heterozygotes, or

(ii) there were in the original population many alleles which slightly lowered fitness in homozygous conditions. Some of these alleles became fixed by chance during the early stages of inbreeding.

It is not easy to distinguish between these experimentally. However, whichever is true, it should be possible ultimately to produce a homozygote of high fitness. Even if there are heterotic loci, sooner or later, as Fisher (1930) pointed out, a duplication of the locus will occur, so that the genotype A/a will be replaced by Aa/Aa; admittedly, the population might have to wait a very long time for the appropriate duplication.

In the present context, the important question is not the detailed mechanism of inbreeding depression, but whether prolonged inbreeding in nature ultimately eliminates it. The evidence is not decisive. Jinks & Mather (1955) argued that hybrid vigour is less marked in habitually selfing species, and this is probably true. But even in strongly inbreeding species, plants produced by outcrossing may have a higher fitness, at least in the F_1. For example, Marshall & Allard (1970) found a higher frequency of adult heterozygotes in wild populations of *Avena sativa* than would be predicted from the frequency in seed.

What is the future of a predominantly selfing species? If a species were wholly selfing, it would suffer the same long-term disadvantages as a parthenogenetic population. However, it seems likely that selfing species will prove to consist of a number of biotypes. This is certainly true of *Avena*, as shown by the work of Allard and Jain and their colleagues. It is also true of the selfing gastropod snail, *Rumina* (Selander & Kaufman, 1973). More important, there is a big difference between 100% and 99% selfing as far as long-term evolutionary results are concerned. Both in *Avena* and *Rumina*, some outcrossing goes on. Theoretical population geneticists have so far had little to say about species with this kind of breeding structure.

B   Resource allocation in hermaphrodites

In this section, I ask in what circumstances a hermaphrodite, by allocating its resources partly to male and partly to female functions, can pass its genes on to a greater total number of offspring than a specialist male or female. I assume that hermaphrodites are self-incompatible, Let m, f and h be the numbers of males, females and hermaphrodites in a population. A single male produces N sperm, a female n eggs, and a hermaphrodite αN sperm and βn eggs. The capacity of a hermaphrodite, compared to a single-sexed individual

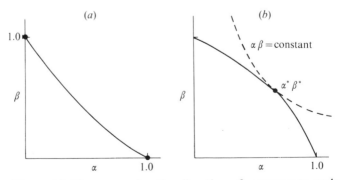

Figure 16. Fitness sets for the allocation of resources to male and female functions. (From Charnov, Maynard Smith & Bull, 1976.) For α and β, see text.

is, therefore, measured by the pair of numbers α, β. It is assumed that the phenotype of an individual is constrained to lie within a 'fitness set' (Levins, 1968), as shown in Figure 16.

I assume initially that the eggs and sperm are shed into a single random mating pool. Let R be the total number of offspring produced by the population. Let W_m, W_f, and W_h be the 'fitnesses' of males, females and hermaphrodites respectively, defined as the expected number of offspring produced in each case. Then

Total sperm $= N(m + \alpha h)$, and

total eggs $= n(f + \beta h)$,

and hence

$$W_m = R\left(\frac{1}{m+\alpha h}\right); \quad W_f = R\left(\frac{1}{f+\beta h}\right);$$

$$W_h = R\left(\frac{\alpha}{m+\alpha h} + \frac{\beta}{f+\beta h}\right). \tag{8.2}$$

If we now suppose that the sex habit of an individual is determined by autosomal genes, then in an equilibrium population the fitnesses of any types actually present must be equal (see Chapter 9, section A, for a discussion of evolutionary stability). Otherwise, a gene converting individuals to the fittest type would increase in frequency.

Consider first a dioecious population, with $h = 0$ and $W_f = W_m$. Then, from equation (8.2), $m = f$. That is, there will be a 1 : 1 sex ratio, as predicted by Fisher (1930); the sex ratio is discussed further in Chapter 9, but note for the present that no concept of parental care has been introduced into the model, so a 1 : 1 ratio is expected. We now ask, could a dioecious population resist invasion by a hermaphrodite mutant? Clearly, it could do so if $W_f = W_m > W_h$, which, from equation (8.2), reduces to

$$\alpha + \beta < 1. \tag{8.3}$$

That is to say, if the fitness set is concave (Figure 16a) then a dioecious population is stable; if the set is convex (Figure 16b), then a hermaphrodite population is stable.

This conclusion was first stated by Charnov *et al.* (1976). That paper also investigated the conditions for gynodioecy and andro-dioecy; it also showed that in a hermaphrodite population the evolutionarily stable resource allocation is that which maximises the product, $a \times \beta$. Note that if egg and sperm production are limited in the main by the same resources, then, approximately, $a + \beta = 1$. If so, $a \times \beta$ is maximised when $a = \beta$; that is, a hermaphrodite should allocate its resources approximately equally to male and female functions (Maynard Smith, 1971*b*).

The important conclusion of this analysis, however, is that a concave fitness set implies dioecy, and a convex set hermaphroditism. The question, of course, is what factors will tend to make the set concave or convex, respectively. Before starting on this discussion, however, it is necessary to relax the assumption that the eggs and sperm are shed into a random mating pool, which was made so that I could assume that all eggs, and all sperm, were equally likely to contribute to the next generation. Notice, however, that the preceding argument holds just as well if a hermaphrodite phenotype (a, β), refers to an individual which is the male parent of a times as many offspring as a typical male, and the female parent of β times as many offspring as a typical female. With this reinterpretation of the model, I now list some factors which will favour a convex set, and hence hermaphroditism; the problem is approached from the point of view of a plant, since most higher plants are hermaphrodite.

(i) In any species, the period during which resources must be allocated to pollen production is earlier than the period when seeds are ripened. Hence a hermaphrodite allocates resources to reproduction over a greater total period than a male or female does. This may mean (although there is no logical necessity) that a hermaphrodite can allocate a greater total quantity of resources to reproduction, and hence a convex set. The same argument applies to animals that brood their young.

(ii) In insect-pollinated plants, it is almost universal for the individual flowers to be hermaphrodite. There is, therefore, a potential saving of resources, since the same organs of attraction serve both the male and female functions. Again, there is no logical necessity that this should lead to a convex set, but for most plausible models of resource allocation it does so; there is room for further research on this topic. If this is an important reason for the prevalence of hermaphroditism in plants, one would expect dioecy to be commoner in wind- and water-pollinated species, and this seems to be the case. In contrast, animal species which have elaborate and expensive structures which serve only or mainly the functions of one sex (e.g. weapons, locomotor organs, intromittent organs) are more likely to be dioecious.

(iii) In sessile organisms, the number of offspring produced per gamete is likely to be a decreasing function of the number of gametes. Consider a hermaphrodite plant. Suppose that if it produces N pollen grains it will pollinate half the ovules of its immediate neighbours; it is unlikely to double this proportion by producing $2N$ pollen grains, and it certainly will not treble it by producing $3N$. The same law of diminishing returns applies to seeds, unless they are few in number or very widely distributed. There is likely to be, between the seedlings of a plant which drops its seeds close to itself, sib competition of a kind which ensures that the fraction of seeds surviving is a decreasing function of the seed number. Either one of these effects will make the fitness set of a hermaphrodite convex. (This argument was pointed out to me by Drs Eric Charnov and James Bull.)

These three effects – a longer period of resource allocation, organs which perform both male and female functions, a law of diminishing returns in sessile organisms because of inter-gamete and inter-offspring competition – are the main ones making for a convex set and for hermaphroditism. An important effect making for dioecy is an increased rate of return with increasing investment. This may happen, for example, in polygamous animals, in which a small male has a low mating success, and a large male a very high one (compare the argument of Trivers & Willard, 1973; see p. 162).

C   Other models for hermaphroditism

Ghiselin (1969, 1974) proposes three models for the evolution of hermaphroditism. These are now briefly discussed:

The low density model

If a species often exists at a low density, or if individuals are sessile or poorly mobile, or if individuals commonly colonise empty habitats, the members of a dioecious species will have difficulty in finding mates. This is one of the major selective factors discussed above as favouring self-compatibility in hermaphrodites; clearly, it would also favour such hermaphrodites over single-sexed individuals. However, even self-incompatible hermaphrodites have some selective advantage in such situations, because every individual a hermaphrodite meets is a potential mate, whereas only half the individuals a single-sex individual meets is a potential mate (Tomlinson, 1966).

It is not easy to tell how important this effect is. As will appear below, hermaphroditism is indeed common in situations where mates are hard to find. In plants, these hermaphrodites are usually self-compatible, and the same is probably true in animals, although data are hard to come by. My own feeling is that there would be strong selection for some degree of self-compatibility, perhaps the degree found in the snail *Biomphalaria*, which can fertilise itself if isolated but which outcrosses if the opportunity exists. However, information on self-fertility of animal hermaphrodites is badly needed. If it turns out that many hermaphrodites in low density situations are self-incompatible, then the effect proposed by Tomlinson may be important.

The size advantage model

This is a model for sequential hermaphroditism, which will be favoured if the male and female functions are best performed by individuals of different sizes. For example, protandrous hermaphroditism will be favoured if egg laying is best performed by a large individual, and protogynous hermaphroditism will be favoured if there is fighting between males for females.

The gene dispersal model

This is not so much one model as a reminder that hermaphroditism has consequences for the genetic structure of populations. Ghiselin (1969) discusses two particular effects. First, sequential hermaphroditism could act as a device to reduce brother–sister mating, just as protandrous and protogynous hermaphroditism tend to reduce selfing in plants. Secondly, hermaphroditism may increase effective population size (Murray, 1964). Thus a population of N hermaphrodites has the same effective population size as one of $N/2$ males and $N/2$ females, and a larger effective size than a dioecious population of N with an unequal sex ratio. It is hard to see how this second effect could be of any selective importance.

In summary, the size advantage and gene dispersal models may have some relevance for the evolution of sequential hermaphroditism. The low density model as formulated by Tomlinson (1966) and Ghiselin (1969) may help to explain why hermaphroditism is found in species in which finding a mate (or achieving pollination) is a problem, but I think that in such species there will be strong selection favouring self-compatibility or parthenogenesis.

D   Hermaphroditism, monoecy and dioecy in plants

The hermaphrodite condition is primitive in higher plants, monoecy and dioecy having arisen independently on many occasions from hermaphrodite ancestors. The advantages of hermaphroditism in terms of resource allocation were outlined in section B of this chapter: resources can be allocated to reproduction over a longer period; the same floral organs can serve male and female functions; in a sessile organism there are diminishing returns from the production of additional gametes of a given sex (the first and last of these advantages accrue also to monoecious plants).

The main selective pressure favouring dioecy appears to have been the avoidance of inbreeding. The same is probably true of monoecy; it seems likely that it is easier for a self-compatible hermaphrodite to evolve monoecy or dioecy than an effective self-sterility mechanism. Two lines of evidence supporting this interpretation are that dioecy tends to occur in families which have not evolved self-sterility (cf. Baker (1959), who points out as an example that the Caryophyllaceae are never self-sterile, but have evolved dioecy on at least three occasions) and that monoecious plants are almost never self-sterile (Godley, 1955).

The basic picture for higher plants, then, is that hermaphroditism is the typical state because of resource allocation advantages; that monoecy and dioecy are more easily accessible but ultimately less efficient adaptations for outcrossing than self-sterility. This picture will be briefly illustrated by island flora, and by the flora of forests.

Baker (1955) argued that most long-distance colonists, for example of oceanic islands, are self-compatible hermaphrodites. This conclusion was queried by Carlquist (1965, 1966), mainly on the basis of his work on the flora of the Hawaiian islands. These islands have 27.5% dioecious species, as compared to values of less than 5% which are typical of continental floras. At first sight this appears to contradict Baker's argument. However, Baker (1967) defended his original position in a brief but stimulating article. In effect, he suggested that the original colonists of Hawaii were indeed self-compatible hermaphrodites, and that dioecy has evolved subsequently. According to this view, the high frequency of dioecy on the islands arose because none of the original colonisers were self-sterile, and species have responded to selection against inbreeding by the relatively easy route of dioecy. Among other arguments in support of this view, Baker points out that there are no heterostylous species on the islands; heterostylous plants are almost always self-sterile, and so would be unable to colonise.

Some data on the breeding systems of forest plants are given in Table 6. The most striking generalisation (already known to Darwin (1876)) which can be made is that there is a much higher frequency of dioecy among trees than shrubs, and among shrubs than herbs. The table shows this only for British woodlands. However, Bawa & Opler (1975), in their study of a tropical forest in Costa Rica, found that 22% of the trees and 11% of the shrubs were dioecious, but only 1 of 300 herbs was, and no climbers or epiphytes were. The obvious explanation is that the larger the plant, the more likely it is to fertilise itself and, therefore, the stronger the selection for a system preventing selfing.

Dioecious trees in tropical forests are in the main insect-pollinated, whereas monoecious and dioecious temperate trees are mainly

Table 6. *Proportions of forest plants with different habits*

Type of forest	Layer	Hermaphrodite	Monoecious	Dioecious	References
Deciduous woodland in Britain (four types)	Tree	50–57	14–30	20–33	Baker (1959)
	Shrub	60–89	0–38	0–23	
	Field	86–93	0–9	0–9	
Five North American temperate forests	Tree	0–27	60–83	6–17	Bawa & Opler (1975)
Seven tropical forests	Tree	41–68	10–22	22–40	Bawa & Opler (1975)

wind-pollinated. Presumably this reflects the relative frequency of insects and of wind in the two habitats. The very high frequency of monoecious trees in North American temperate forests is puzzling. Monoecy in a tree can be a reasonably effective outcrossing device, since male and female flowers are usually borne on different parts of the tree. Monoecious plants are usually wind- or water-pollinated; they, therefore, lack attractor organs such as petals and nectaries and would gain nothing in resource allocation terms by having hermaphrodite flowers. Thus if a plant lives in a habitat in which wind is an adequate pollinator, monoecy ensures a degree of outcrossing combined with an efficient allocation of resources.

Finally, the data provide only weak support for the view that selection for outcrossing is stronger in tropical than in temperate habitats (see p. 109). A range of 22–44% dioecy among tropical trees is very high, but this seems mainly to reflect the fact that they are trees, and not that they are tropical.

E   Hermaphroditism in animals

The best reviews of hermaphroditism in animals are by Ghiselin (1969, 1974). The main ecological and life history correlates of hermaphroditism are as follows:

(i) The sessile habit; for example most endoprocts, brachiopods, and crinoids. There are, however, many exceptions.

(ii) Parasitism; for example complete taxa, such as flukes and tapeworms, and many isolated species in primarily free-living gono-choristic groups, such as crustaceans, nemertines, and prosobranchs. The main exceptions are the Nematoda and Acanthocephala.

(iii) The deep sea; for example among bony fish, bivalves, and crustaceans.

(iv) Planktonic; for example salps, chaetognaths and pteropods.

(v) Brooding; hermaphroditic species which brood their young are found in a number of groups, including sea anemones, bivalves, brachiopods, and echinoderms.

It is hard to say how far these associations are to be explained by the advantage in finding mates and how far by resource allocation.

The prevalence among deep-sea and parasitic organisms tends to support the former selective mechanism. The association with brooding, which is most convincing, is most easily explained by resource allocation; an animal which broods its young allocates resources over a longer period, just as a hermaphrodite plant does. The association with a sessile habit could be explained by either process. However, many sessile organisms are so abundant that it is hard to believe that finding a potential mate is a serious problem. As against this, many sessile animals have planktonic larvae, so that one of the arguments from resource allocation (the law of diminishing returns from producing additional eggs) does not apply. Finally, the association with the planktonic habit is something of a puzzle, although it may be that the resource allocation arguments for sessile organisms apply also to planktonic ones.

I find it hard to assess these comparative data. A major difficulty is that whether hermaphrodites are self-fertile is only rarely recorded. At least some helminth parasites are known to be so; the degree of selection arising because a single individual can colonise a host must be substantial. But there is need for further investigation, carried out with a clear idea of the alternative selective possibilities.

F   Avoidance of inbreeding in animals

There is likely to be a balance of selective forces for and against inbreeding in gonochoristic animals, parallel to the balance between selfing and outcrossing in hermaphrodites. The low fitness of inbred individuals will select for outcrossing, but this must be balanced against selection for inbreeding for reasons similar to those which favour selfing in hermaphrodites. Thus, if there were no deleterious effects of inbreeding, a female who mated with her brother would increase her inclusive fitness by increasing the number of her brother's genes passed to future generations (Bengtsson, 1978; this assumes that there is little or no paternal care, and that a male who mates with his sister does not thereby diminish his chances of mating with other females).

In general, the balance of selective forces seems to have favoured outcrossing. There are two ways in which an animal might reduce the frequency with which it mates with a close relative:

(i) By recognising relatives, and refraining from mating with them.

(ii) By dispersing before sexual maturity in such a way that close relatives are unlikely to be encountered.

Recognition could depend on direct recognition of a genotypic similarity which indicated relationship. I know of no case in higher animals in which inbreeding is prevented by such a mechanism, although mating types in Protista and self-incompatibility alleles in higher plants are mechanisms of this kind. One case in higher animals in which mating propensity may be influenced by recognition of genotype is the 'rare male phenomenon' in Diptera (Ehrman, 1972), but here females select males which are of a locally rare type, rather than unrelated. However, if the rare male phenomenon depends on the recognition of genotypes, we may yet find cases in which inbreeding is avoided by genetic recognition of relations.

It seems more likely, however, that animals would, for example, treat as a sib all others raised by the same parents, whether or not they were genetic sibs. This behaviour could evolve more easily, and in nature it would have the same effect of reducing inbreeding. There is some evidence that this does happen. Hill (1974) showed that in the deermouse, *Peromyscus maniculatus,* there is delayed reproduction if the two potential parents have been reared together, whether or not they are actual sibs. The delay is probably caused by delayed mating, but may be caused by failure at later stages, for example implantation. As one would expect, Hill also found smaller litters, and higher mortality, from matings of true sibs than from those of foster sibs.

The effect observed by Hill may be of widespread occurrence, but probably the main way in which inbreeding is avoided is through differences in dispersal patterns between the sexes. In some cases it may be impossible to distinguish the two causes, since dispersal of members of one sex may be caused by the refusal of the other sex to mate with them.

In birds, the avoidance of inbreeding seems to depend more on patterns of dispersal than on recognition of relatives. Greenwood & Harvey (1976, 1977) showed that in seven passerine species which breed in a feeding territory established and predominantly defended by the male, individuals of both sexes usually return to breed close to where they were fledged, but breeding site fidelity is greater in the males, the females being more likely to breed at a substantial distance from their birthplace. They believe that this asymmetry evolved because of the deleterious effects of inbreeding. Clearly the opposite asymmetry would serve equally well; they suggest that the sex which establishes the territory gains an additional advantage from doing so in a familiar place. Their general interpretation is supported by the fact that there is no sex difference in dispersal in the green finch, *Carduelis chloris,* which shows a high level of dispersal because of an unpredictable food supply. Each pair holds a small non-feeding territory, and the birds pair in winter flocks before selecting a breeding site.

There is at present no reason to think that there is any additional barrier to inbreeding in birds, based on recognition of relatives. Bulmer (1973) found three cases of brother–sister mating and two mother–son matings in the data on great tits (*Parus major*) from Wytham Wood near Oxford. These numbers correspond rather closely to the number expected from the known dispersal patterns of the birds, assuming no avoidance based on individual recognition. He did find some evidence of inbreeding depression, both in nest mortality, and in survival to breed. In fact, of forty-two birds fledged from these five pairs (seven broods), none survived to breed. This is not as striking as it sounds; the expected number of survivors is approximately three.

In group-living mammals it is common for males raised in one group to transfer to another before breeding. The olive baboon, *Papio anubis,* affords a striking example. Packer (1975) found, in a six-year study of three troops, that of forty-one known transfers into or out of a troop, thirty-nine were males, the other two being a female and her immature daughter. Of the males which reached maturity during the study period none remained in its natal troop. To breed, a male must leave his natal troop and fight his way into another. There are indications that an immediate reason why males leave their natal troop when they become sexually mature is that the females in that troop refuse to mate with them.

There is a rapidly growing literature on this topic, particularly for mammals. Male transfer seems to be the rule. For example, Drickamer & Vessey (1973) described a pattern of male transfer in the Indian rhesus monkey, *Macaca mulatta,* introduced on islands in Puerto Rico, similar to that later observed by Packer in the olive baboon. However, female transfer is known to occur in the chimpanzee (Pusey, 1977), the gorilla (Harcourt, Stewart & Fossey, 1976), and the South African hunting dog, *Lycaon pictus* (Frame & Frame, 1976).

It is hard to see why these inter-group transfers should have evolved, unless it was because of the deleterious effects of inbreeding. As pointed out above, if it were not for these effects, a female could increase her inclusive fitness by mating with her brothers rather than by rejecting them. As yet, we do not know of any mammalian groups which are predominantly endogamous, but it will be surprising if none is found.

G   The human incest taboo

It is one of the few universal features of human societies that certain types of marriage between close relatives are proscribed. Marriages between first degree relatives (parent–offspring, sib–sib) are forbidden almost always, and most societies have additional prohibition on more distant relatives (e.g. cousins). There are three possible origins of these prohibitions:

(i) The origin and maintenance of these prohibitions has nothing to do with the biological effects of inbreeding. Their function must be sought elsewhere, for example as aids to economic exchange or alliances between groups. This appears to be the majority view of anthropologists.

(ii) Incest is avoided because people observed the adverse effects of inbreeding, correctly deduced the causal connection, and proscribed incest in order to prevent the undesirable biological consequences. This view seems implausible. Schull & Neel (1965) found from a study of cousin marriages in Japan that inbreeding depression does occur in man, but to do so they had to use the full armoury of modern statistical techniques. The effect of father–daughter and sib–sib mating is, however, more extreme (Adams & Neel, 1967) and could conceivably have been observed in early times. If so, the deleterious effects of cousin marriage might have been inferred by extension. But the process does not seem likely.

(iii) Our pre-human ancestors behaved in ways which reduced inbreeding, because such behaviour was naturally selected. When symbolic communication, myth-making and cultural transmission became the dominant modes of communication, these earlier 'incest barriers' changed into 'incest taboos'.

It seems to me entirely plausible to suppose that our animal ancestors evolved barriers against incest, because other primates are known to have done so. It also seems likely that when our ancestors became self-conscious, they invented reasons, myths and social sanctions to reinforce what they were already doing. When Levi-Strauss (1968) asserts that the incest taboo is the characteristic feature which originated human culture, it is in one sense tautologous and in another manifestly false. If the emphasis is on the cultural connotations of the word 'taboo', the statement is tautologous since there can be no culture without culture. If the statement implies that animals do not have practices which avoid incest, then the statement is false.

Is there any direct evidence that incest avoidance in man is more than a cultural invention? Perhaps the most striking evidence comes from the study of Israeli kibbutzim, in which children of both sexes are brought up communally (Talmon, 1964; Shepher, 1971). Although there is no conscious social pressure against marriages between members of the same peer group, Shepher reports that there was not one case among 2769 marriages between second generation kibbutz adults. The implication is very strong that, when seeking sexual partners, human beings seek for preference someone not closely familiar to them as a child. Such a preference could well have evolved because it reduced inbreeding.

One argument which has often been used against a biological interpretation of the origin of incest avoidance in man is that the details of the prohibitions differ between societies in ways which are not always consistent with degrees of genetic relatedness. It is also pointed out that marriage rules of a society may change far more rapidly than could be accounted for by genetic change.

These objections have recently been discussed by Alexander (1977a). He accepts unreservedly that differences between cultural patterns in different societies are not caused by genetic differences between their members. He then goes on to point out that some features of human kinship systems which have been quoted as being inconsistent with the view that human beings behave so as to maximise their inclusive fitness are in fact quite consistent with that view, once the uncertainty of paternity is allowed for. There are two examples. The first is that in many societies a child's mother's brother may be as important, or more important, in providing support as the child's putative father. If paternity were certain this would be inconsistent with maximising inclusive fitness. If paternity is uncertain, the putative father may have no genes in common with the child, whereas the mother's brother must have at least $\frac{1}{8}$ of his genes in common. However, as Greene (1978) has pointed out, a man's chance of being the father of his wife's children must be less than 27% before the mother's brother will have, on average, more genes in common with the child than the mother's husband.

The second case concerns prohibitions against marrying different categories of cousin, so called 'parallel' cousins in which the relationship is through two sisters or two brothers, and 'cross' cousins, related through a brother and sister. Most modern societies do not distinguish between these categories, and with enforced monogamy the degree of relatedness is the same. In primitive societies Alexander points out that it will often be the case that parallel cousins are more closely related; this will be the case, for example, if a man successively marries two sisters, or if a man hands on a wife to a brother. The relevance of this is that prohibition of marriage between parallel cousins is commoner than between cross cousins.

At this point, Alexander (1977b) recognises that there is no likelihood that argument based on inclusive fitness will be listened to by sociologists or anthropologists unless a proximate mechanism can be suggested; that is to say, until we can suggest a way in which the appropriate behaviour could appear during individual development.

Such a mechanism would have to be consistent with what is known about human learning. Thus he accepts that the different patterns of behaviour appropriate to different societies do not evolve genetically as adaptations to those societies. What is needed is a developmental process dependent upon a genetic constitution which does not differ between societies, which ensures that individuals will acquire patterns of behaviour which are different in different societies but nevertheless are such as to maximise fitness in those societies.

I do not think it will be easy to suggest such a proximate mechanism. It seems to me more plausible that the differences between the customs of different societies call for a cultural interpretation which has little to do with inclusive fitness. But to attempt a general explanation of human marriage customs and incest taboos which ignores the fact that, in all probability, our ancestors avoided mating with close relatives long before they could talk would be foolishly parochial.

9 *Anisogamy and the sex ratio*

A   Methods: evolutionarily stable strategies

In this chapter I take it for granted that meiosis followed by syngamy is a necessary feature of the life cycle, and ask why and in what circumstances it is selectively advantageous to produce gametes of different sizes (anisogamy) and, given that different gamete sizes are produced, what determines the frequency in the population of individuals producing large and small gametes (the sex ratio). Work on the evolution of the sex ratio, from Fisher to Hamilton and Trivers, forms one of the most satisfying threads in evolution theory. The ideas are elegant, and are often testable. They are, however, unexpectedly difficult, so that a few preliminary remarks may be helpful.

The essential concept is that we should concentrate on a gene which can influence the sex ratio, and ask two questions: first, how does the gene affect sex ratio; secondly, how can the gene maximise its representation in future generations? I will consider these questions in turn.

How does the gene affect the sex ratio?

(i) A gene may act in a reproducing individual so as to alter the proportions of the two sexes among its offspring. It could do this in the heterogametic sex by altering the ratio of male- and female-determining gametes produced; in the homogametic sex it could alter the success in fertilising of male- and female-determining gametes; in either sex it could alter the survival chances of offspring of the two sexes. But in all cases it would act by influencing the 'behaviour' of the parent.

(ii) A gene could influence the sex of the individual in which it finds itself.

(iii) A gene could influence the success of the chromosome on which it finds itself during meiosis (i.e. 'meiotic drive') or of the gamete in which it finds itself. These two processes, although physiologically different, are grouped together because they may be difficult to distinguish observationally.

In all the above cases, although the genes act in the parent, the offspring, and the gamete, respectively, they are transmitted chromosomally. A further category of gene is one which is transmitted cytoplasmically, and which influences the sex of the individual. This last category I shall not consider further. However, cytoplasmic 'genes' are known to occur (e.g. the 'sex ratio' condition in *Drosophila* (Poulsen & Sakaguki, 1961); male sterility in gynodioecious plants (Lewis, 1941)).

Most discussions of the sex ratio, including Fisher's (1930), consider only genes of type (i) above. Difficulties arise when genes of more than one type exist simultaneously. For example, when Trivers & Willard (1973) discuss 'parent–offspring' conflict over the sex ratio, they are in effect discussing a conflict between genes of types (i) and (ii). I shall first discuss the sex ratio on the assumption that only genes of type (i) exist, and then some of complications which arise because genes act at other phases of the life cycle.

How can one decide how the sex ratio will evolve? One method would be to consider a pair of alleles, say *A* and *a,* with different effects on the sex ratio, and to ask whether or not one or other will go to fixation, and if not, what will be their equilibrium frequencies. However, it is more effective to seek for an 'evolutionarily stable strategy', or ESS. This approach is implicit in Fisher's (1930) book. Hamilton (1967), writing after the development of the theory of games by Von Neumann & Morgenstern (1953), explicitly thought of an 'unbeatable strategy'; a similar mathematical approach, but without an explicit reference to game theory, had been used by MacArthur (1965). The concept of an ESS, which is similar to but more general than Hamilton's 'unbeatable strategy', was developed independently by Maynard Smith & Price (1973) to analyse contest strategies in animals, and has since been applied to a number of evolutionary problems in which the optimal strategy for an individual to adopt depends on what others are doing (Maynard Smith, 1976e).

In colloquial terms, an ESS is a strategy such that, if almost everyone is doing it, it would not pay a mutant to do something different. More formally, and in the context of the sex ratio, we proceed as follows. We seek for a sex ratio s^* which is evolutionarily stable in the following sense: given that the population sex ratio is in fact s^*, then no mutant allele g which produces a sex ratio s different from s^* is as successful (in the sense of producing as many copies of itself in future generations) as the typical allele g^* which produces the sex ratio s^*. If we can find a sex ratio s^* with this property, then the population will evolve until it has that stable ratio, provided of course that appropriate genetic variation exists.

B   Genetic variance of the sex ratio

One further point must be discussed before we come to grips with the problem. Surely, one might argue, the sex ratio is $1:1$ because sex is determined by genes which segregate in a $1:1$ ratio in meiosis. There is some force to this objection. Even if genes at two independent loci determine sex, it can be shown (Scudo, 1964) that, with a regular $1:1$ meiotic segregation, there is only one way of achieving a population sex ratio different from $1:1$ (double heterozygotes A/a, B/b one sex, all other genotypes the other); all other mechanisms give rise to a $1:1$ ratio.

The answer to this objection is that there are various ways in which genes might alter meiosis or distort its consequences. However, there is little direct evidence for the existence of such genes.

Probably the most extensive data are for the human species; they have been analysed by Edwards (1962, 1966, 1970). There are some serious statistical traps. It is not satisfactory to examine, for example, all families of three children, and ask whether they fit the binomial expectation, because of voluntary family limitation; for example, a couple might stop after three children if they have had at least one child of each sex, but continue if they had not. For this reason it is necessary to know the birth order. When this is taken into account, it turns out that there is an effect of parity, the later children being more likely to be female (Novitski & Kimball, 1958; Teitelbaum, 1972). Edwards found that there was a positive correlation between the sexes of successive children in the same family, but when he compared the sexes of two children in the same family, separated by one or more births, there was no association. He interprets the association between successive children as a temporary environmental effect, and concludes that there is no evidence of between-family variation in the sex ratio, and hence that there is no variation which could be genetic.

Falconer (1954) drew a similar conclusion from his experimental study of mice and *Drosophila*. In both species he practised between-family selection; i.e. in each generation he set up a number of paired matings, estimated the sex ratio from each, and selected males and females as parents of the next generation from those families whose sex ratio showed the greatest bias in the desired direction. The response to selection was not statistically significant in either direction either in mice or in the main *Drosophila* experiment (although in a preliminary experiment he did get a response to selection for an excess of females, which he believes to have been because of the presence of a recessive sex-linked lethal in the foundation population). An analysis of heterogeneity between families also failed to reveal any statistically significant variability in either species.

Falconer does point out that his experiments do not exclude the possibility that there is sufficient genetic variability for the sex ratio to enable natural selection to alter it in wild populations. His experiment suffered from the unavoidable difficulty that any measurement of the sex ratio in a family is subject to large sampling error, which may have been sufficient to swamp any real differences. It is therefore interesting that Weir (1953) was able to produce a striking difference in the sex ratio of mice by selecting, not for the sex ratio itself, but for blood pH. If two characters A and B are correlated, but A can only be measured with an enormous sampling error, it may be easier to alter A by selecting for B. However, the correlation between blood pH and sex ratio in Weir's experiment seems to have been accidental, perhaps because of linkage disequilibrium in the foundation population. When Weir & Clark (1955) repeated the experiment, they again obtained a response in blood pH, but this time there was no associated change in sex ratio. In the first experiment, it was the males which were responsible for the changed ratio, probably by producing a changed ratio of X- and Y-bearing sperm (Weir, 1962).

Particularly favourable data on the genetics of the sex ratio are provided by domestic cattle. Bar-Anon & Robertson (1975) analysed over 150 000 births sired by 107 bulls. They found significant differences between bulls, the standard deviation in the proportion of males being 1.5% (after correcting for the expected binomial variation). There was also a significant correlation of 0.5 between the sex ratio from a bull and from the father of that bull, when known. It seems clear that genes are influencing the sex ratio, but it is not clear whether the differences are caused by autosomal genes active in the male parent, by Y-linked genes with meiotic drive, by Y-linked genes influencing male survival, or by autosomal genes influencing the survival of male and female offspring differentially.

Thus, there is little direct evidence for the existence of genetic variance of the sex ratio. The main reason for supposing such variance to exist is that there are known to be a number of environmental factors (e.g. the parity effect described above) which do affect the sex ratio, and it is a good commonsense principle that if environmental factors can affect some characteristic, it is likely that genes will do so also.

These reservations concerning the extent of genetic variation in the sex ratio do not apply to haplo-diploid species, in particular the Hymenoptera. In such species it is almost always the case that mated females lay both fertilised (female) and unfertilised (male) eggs, and there can be little doubt that genes acting in the female could alter the sex ratio among her offspring. Similar considerations apply in the less common cases in which sex is determined before fertilisation. For example, in the worm *Dinophilus* the female produces large and small eggs which, after fertilisation, develop into females and males respectively (Bacci, 1965). There are strain differences in sex ratio; the species would provide an admirable test of Fisher's investment theory of the sex ratio were it not that parthenogenetic reproduction apparently also occurs.

C   Anisogamy

Why do higher organisms, plant and animal, produce gametes of two different sizes? What I believe to be essentially the correct answer to this question was given by Parker, Baker & Smith (1972). Here I present their argument in a more general form; the approach was suggested by Bell (1978).

Consider a population of organisms each of which is capable of producing a mass M of gametes, where M is a constant. Genotypes differ in the mass per gamete, and the number produced, subject to the constraint

$$nm = M, \tag{9.1}$$

where n is the number of gametes, and m the mass of each (if a cell produced 'gametes' by binary fission, then $n = 2$, $m = M/2$, and M is the mass of the cell before fission).

I assume at present that gametes fuse randomly in pairs to form zygotes. If a and b are the masses of the fusing gametes, the zygote has mass $a + b$, and has a probability of surviving to become an adult of $S_{a+b} = \phi(a+b)$. That is, the probability of survival depends on total mass, and not on the relative contributions of the two parents. The mass of the smallest cell which can function as a gamete is taken as δ.

Subject to fixed values of δ, and of S as a function of $(a+b)$, will isogamy or anisogamy be an ESS, and what will be the stable gamete sizes and sex ratio?

Consider first a population producing gametes of size δ, and ask, is such a population stable against invasion by a mutant producing gametes of size $m \gg \delta$? Measuring the fitnesses, W_δ and W_m of typical and mutant individuals by the number of surviving zygotes to which they contribute gametes, we have

$$W_\delta = \frac{M}{\delta} S_{2\delta}; \quad W_m = \frac{M}{m} S_{m+\delta}.$$

Then the population producing microgametes is stable, provided that $S_{2\delta}/\delta > S_{m+\delta}/m$; this condition is illustrated in Figure 17a.

I believe that the condition in Figure 17a is primitive for eukaryotes; i.e. the earliest sexual eukaryotes were isogamous, and

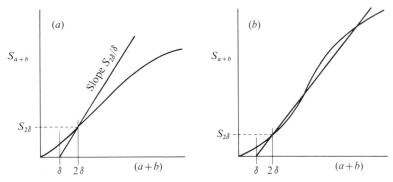

Figure 17. Zygote survival, S_{a+b}, as a function of gamete size, $a+b$. (*a*) Production of microgametes is evolutionarily stable; (*b*) production of microgametes is not stable. For other symbols, see text.

produced gametes as small as was compatible with effective gamete function. In a small unicellular eukaryote, this may well have amounted to the production of four isogametes from a typical diploid cell, without further size reduction.

With increasing size of the adult organism, particularly if the adult were multicellular, the chance of a zygote formed by the fusion of two microgametes surviving to the adult stage may have become very small, leading to the condition shown in Figure 17*b*. At this stage, a population of microgamete-producers could be invaded by a new macrogamete-producer. The question then arises, would the new type replace the δ-producers, or would anisogamy be stable?

I first seek a gamete size $m^* \gg \delta$, such that m^* is stable against invasion by gametes slightly larger or smaller, and then ask whether a population of m^*-producers could be invaded by a δ-producer.

Consider a mutant producing gametes of size m in an m^* population. Then

$$W_m = \frac{M}{m} S_{m^* + m},$$

and

$$\frac{\partial W_m}{\partial m} = \frac{M}{m}\left(\frac{\partial S_{m^* + m}}{\partial m}\right) - \frac{M}{m^2} S_{m^* + m}.$$

If m^* is an ESS, then $\partial W_m/\partial m = 0$ when $m = m^*$, which gives

$$(\partial S_{m^*+m}/\partial m)^* = S_{2m^*}/m^*. \qquad (9.2)$$

This condition is shown graphically in Figure 18. Provided that the slope of S_{m+m^*} *vs* $m+m^*$ is at some point greater than $2S_{m+m^*}/(m+m^*)$, there will be a locally stable gamete size m^*.

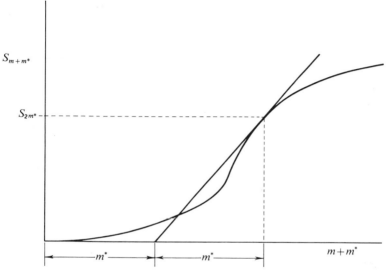

Figure 18. Conditions under which the production of gametes of size m^* is evolutionarily stable. For other symbols, see text.

But will a population producing gametes of size m^* be stable against invasion by a mutant producing gametes of size δ? There are two possible situations, shown in Figure 19a, b. The fitness of a typical individual is $W_{m^*} = MS_{2m^*}/m^*$, and the fitness of a δ-producer is $W_\delta = MS_{m^*+\delta}/\delta$. Thus in case (a), $W_\delta > W_{m^*}$, and a δ-producer can invade. Case (b) illustrates the conditions which must be met if isogamy is to be stable. They are very severe, since they imply that if $2m^*$ is the optimal mass for a zygote, then a zygote of half that mass has almost zero probability of survival.

I doubt whether the situation in Figure 19b has ever existed; certainly it cannot have been primitive. That is, primitive isogamous

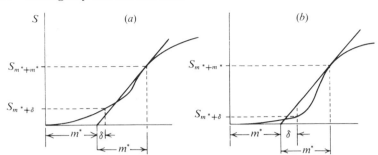

Figure 19. Survival curves for which macrogamete production is unstable (*a*), and stable (*b*). Symbols as for Figures 17 and 18.

organisms should be seen as microgamete-producers ('males'), with macrogamete-producers ('females') a later evolutionary development (Genesis, 2:22).

Given that a population of δ-producers is unstable, and is invaded by individuals producing macrogametes, what will be the result? Let $\hat{m}$ be the evolutionarily stable size of the large gametes. For simplicity I assume that $m \gg \delta$, so that m gametes typically fuse with δ gametes. Then $W_m = MS_{m+\delta}/m$, and the conditions that W_m be at a local maximum when $m = \hat{m}$ gives

$$(\partial S_{m+\delta}/\partial m) = S_{\hat{m}+\delta}/\hat{m} \simeq S_{\hat{m}}/\hat{m}. \tag{9.3}$$

A geometrical interpretation is given in Figure 20. Once a population of δ-producers has been invaded by macrogamete-producers, it will pay to produce microgametes which fuse only with macrogametes and not with other microgametes. If the sex ratio is P δ-producers to $(1-P)$ $\hat{m}$-producers, then the numbers of gametes are in the ratio P/δ to $(1-P)/\hat{m}$, and if all $\hat{m}$ gametes fuse, a fraction $(1-P)\delta/P\hat{m}$ of δ gametes fuse. Hence $W_{\hat{m}} = MS_{\hat{m}+\delta}/\hat{m}$ and $W_\delta = MS_{\hat{m}+\delta}(1-P)\delta/P\hat{m}\delta$. The fitnesses of the two types are equal only when $P = 1 - P$, or $P = \frac{1}{2}$; that is, the stable sex ratio is 1 : 1. It is not difficult to see from Figure 20 that no third 'sex' could invade.

There remains the question of why macrogametes do not evolve the capacity to fuse with other macrogametes, since this would increase their fitness (provided $S_{2\hat{m}} > S_{\hat{m}}$). This problem has been discussed by Parker (1978), who suggests that (1) the risk of failing to find a

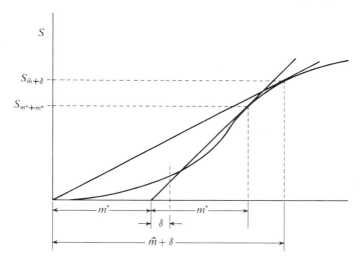

Figure 20. Evolution of anisogamy. For symbols, see text.

partner to fuse with, (2) the cost to a macrogamete of retaining motility, and (3) the risk of inbreeding are sufficient to explain why disassortative fusion has become the rule.

How far does the theory explain the facts? Since the facts were familiar before the theory was developed, it would be odd if the agreement were not reasonably good. The basic conclusion is that the primitive condition is the production of microgametes of the smallest practicable size, and that anisogamy will evolve, through invasion by macrogamete-producers, when adult size is such that it is difficult or impossible for a single motile cell to grow and differentiate into the adult form. The green algae are the group which could best be used to test the model. Knowlton (1974) pointed out that the model of Parker *et al.* (1972) is supported by the Volvocidae, in which the smallest colonies are isogamous, medium-sized colonies show slight anisogamy with no disassortative fusion, and the largest colonies are anisogamous with disassortative fusion. But further work is needed to test the model presented here.

The conclusion that there is a single optimal zygote size is one which would follow from almost any model; indeed it hardly needs a model. How far is it true that members of a given species produce eggs or seeds of constant size? Seed size has been admirably reviewed by Harper, Lovell & Moore (1970). In many species, seed size is so constant that seeds have been used as a primitive unit of weight. Wheat grown at different densities showed a 833-fold variation in the number of seeds per plant, but only a 0.04-fold variation in the mean weight of seeds. Exceptions to this rule of constant seed size tend to occur in species in which there is a protracted period of determinate growth after the time when the number of seeds has been determined and before their final size is reached. It is perhaps better to regard this variation more as unavoidable than as adaptive.

It is not uncommon for an individual plant to produce two sharply distinct types of seed, the determination being developmental rather than genetic. In these cases, the two types of seed usually differ in dormancy and powers of dispersal, and are adapted to different ecological niches. Less commonly, as for example in the corn spurrey, *Spergula arvensis,* there is a genetic polymorphism in seed properties (Harper *et al.,* 1970).

Plants of a given species might be expected to produce larger seeds when intra- or inter-specific competition between seedlings is likely to be high. I know of no evidence that individual plants can achieve such an adjustment of seed size through developmental flexibility, but different populations may show an appropriate genetic adaptation. For example, Werner & Platt (1976) studied populations of six species of goldenrod, *Solidago,* from a prairie and an old-field habitat. They found, in all the species, that seeds from the old-field habitat, in which seedling competition is less extreme, were smaller, and had higher powers of dispersal.

Data on egg size in animals are less extensive. In general, animals seem to resemble plants in that egg number is much more variable than egg size. However, it is often the case in fish that large females lay larger eggs than small ones do (Nikolskii, 1969), and the same is true of some reptiles (e.g. Pimentel, 1959) and amphibians (e.g. McAlister, 1962). The reason for this is far from clear. One possible explanation is that eggs laid by large females are more likely to be exposed to severe intraspecific competition, but it is not clear whether this will always be so.

D   The sex ratio with random mating

Fisher's (1930) reasons for thinking that a $1:1$ sex ratio would be evolutionarily stable can be understood with the minimum of mathematical argument. Thus, suppose a majority of the children being born at any time are females. Then a parent who produces only sons will, on average, have more grandchildren than one who produces only daughters, or a mixture of sons and daughters. Therefore, when there is an excess of females in the population, genes tending to restore the ratio towards unity will spread; the same is, of course, true if there is an excess of males.

Notice that genes in the homogametic parent, autosomal genes in the heterogametic parent, and genes determining the sex of the individual will all tend to produce a $1:1$ ratio. The only genes distorting sex ratio which would be selected are those on the X and Y chromosomes of the heterogametic sex.

In fact, Fisher argued that it was not the number of males and females which would be equal, but the parental expenditure on males and females. Thus suppose that a single parent can produce m sons and f daughters, counted at the end of parental care. The litter (m, f) is constrained to lie within a fitness set. If sons and daughters are equivalent, the boundary of the set is $m+f=$ constant; if a son costs α times as much as a daughter, then $\alpha m+f-$ constant; if sons and daughters utilise in part different resources, the set may be convex. Given random mating, what sex ratio (m^*, f^*) is evolutionarily stable? It is shown in the appendix to this chapter that the stable sex ratio is that which maximises $m^* \times f^*$; i.e. the product of the number of sons and daughters.

If the costs of males and females are equivalent, then $m+f=$ constant, and $m \times f$ is maximised when $m=f$; that is, the stable sex ratio is $1:1$.

If males cost α times as much as females, then $\alpha m+f=$ constant, and $m \times f$ is maximised when $f+m(df/dm)=f-\alpha m=0$, or $f=\alpha m$. That is, at the stable sex ratio, expenditure on females, kf, equals expenditure on males, $k\alpha m$. This is Fisher's (1930) conclusion that expenditure on males and females should be equal.

One point needs discussion. Why do we assert that $m \times f$ is maximised at the end of parental care? There are two ways of seeing the answer to this question. The first is to appreciate that we are maximising W_m subject to a constraint, which is set by what the parents are able to raise. Alternatively, we can suppose that the chances of males and females surviving to breed, given that they were alive at the end of parental care, are S_m and S_f respectively. The breeding population then consists of $S_m \times m^*$ males and $S_f \times f^*$ females. Since S_m and S_f are constant, a sex ratio which maximises $m^* \times s^*$ will also maximise $S_m \times m^* \times S_f \times f^*$. Thus, it makes no difference whether we consider the sex ratio at the end of parental care or at breeding; we draw the same conclusions.

If offspring of one sex, let us say males, are more likely to die before the end of parental care, then a parent who is to invest equally in the two sexes must produce, at conception, an excess of males. This is because, at conception, males are going to cost less than females. If we suppose that, apart from mortality, males and females require the same investment, then if the males die before any expenditure on them has been made (an impossible hypothesis), the primary sex ratio would be adjusted in such a way that the sex ratio after mortality would be 1:1. If males die after some expenditure has been made, there would be an excess of males at conception, but an excess of females after mortality has acted. These statements, of course, assume that the sex ratio is optimised by selection on genes acting in the parent.

Matters become still more complex in an age-structured population. This problem has been analysed by Charlesworth (1977). With human populations in mind, he assumes that male mortality is higher than female, and that offspring that die can be replaced at less cost than offspring that survive to the end of parental care. One interesting result of his analysis is that it is evolutionarily stable for a female to shift the sex ratio of her offspring in favour of daughters as she grows older. This conclusion can be partly explained by the following intuitive argument: a young female can replace a lost foetus, and should, therefore (as argued in the last paragraph), bias her sex ratio in favour of sons; an older female is likely to die or become sterile before she can replace a lost offspring, and should therefore produce a primary sex ratio of unity.

The satisfying feature of this result is that it corresponds rather well with the facts. In our species, older mothers do produce a higher proportion of females. In fact it is probably birth order rather than maternal age which is the proximate cause (Novitski & Kimball, 1958; Teitelbaum, 1972), but this does not alter the argument. A less satisfying feature is that the same facts can be explained in a different way, as will emerge later when discussing the arguments of Trivers & Willard (1973).

A test of Fisher's expenditure theory requires that we look at sex ratios when males and females do in fact have different costs. In those mammals in which breeding males are substantially larger than females, the sizes at weaning are approximately equal. Consequently we would expect a $1:1$ ratio at weaning, even if male mortality is subsequently much higher. In some birds, however, there is a sex difference in size at fledging. Howe (1977) has looked at sex ratios in the common grackle, *Quiscalus quiscula,* in which males are 20% heavier at weaning. He found $52:83$ males to females at fledging in Michigan populations. The difference appears to arise because of a heavier male mortality, since he observes a $1:1$ ratio among embryos.

The most extensive investigation of the sex ratio in a sexually dimorphic bird is by Newton & Marquiss (1978) in the European sparrowhawk, *Accipiter nisus.* In this species, adult females are twice as heavy as males. They hatch from similar eggs, but a twofold weight difference exists already at fledging. The authors sexed 651 broods just before fledging, and found 1102 males and 1061 females. The sex ratio was the same in broods in which all the eggs laid survived to fledge as it was in broods with mortality.

At first sight these data appear to contradict Fisher's prediction that the parents should equalise investment rather than numbers. However, male and female nestlings ate the same amount. Females put on weight faster, but males feather sooner and leave the nest three to four days earlier. Hence parental investment in males and females is equal, at least until fledging (the young are fed for two to three weeks after leaving the nest, but no data are available on food intake during this period). Clearly, it would be dangerous to assume in other species that parents invest differently in offspring of the two sexes merely because they are of different sizes.

In a sense these results, although clear, are disappointing. They are consistent with Fisher's theory, but fail to provide a strong confirmation of it, as would have been the case if a distorted sex ratio had been associated with a different investment per individual male and female offspring. After all, a 1 : 1 sex ratio can equally well be explained by saying that that is what meiosis produces.

E   The sex ratio with local mate competition

As Hamilton (1967) was the first to appreciate, Fisher's conclusion breaks down if there is local competition for mates. Thus consider the extreme case of a population which reproduces only by brother–sister mating. If females lay a fixed number of eggs, and if a single male can mate all his sisters, then a parent which produces only one male will produce more grandchildren than one which produces equal numbers of males and females.

Hamilton lists some twenty-five species of mites and insects, from sixteen families, which combine habitual inbreeding, a large excess of females, and an arrhenotokous breeding system. The relevance of arrhenotoky is that a female can produce a batch of progeny containing only one male by fertilising all the eggs she lays except one. The extreme example is a mite, *Acarophenax,* in which the female produces live young; the single male in the litter hatches, mates with his fifteen or so sisters and dies before he is born.

It is less clear whether a similar association between inbreeding and an excess of females can evolve in the absence of arrhenotoky. In plants, close inbreeding is usually associated with hermaphroditism.

What is the stable sex ratio if inbreeding is only partial? Thus suppose that in every generation a fraction k of all females are mated by their brothers, and $1-k$ mate randomly. Let the population sex ratio be s^* males to $1-s^*$ females. Each family is of size r, and the whole population is of size N. Consider a female GM ('GM' because she will be considered in her role as a grandmother; an exactly similar argument could start from a potential grandfather) carrying an autosomal gene causing her to produce s males to $1-s$ females. Let T be the number of copies of this gene which she transmits to her grandchildren.

To find T we proceed as follows. The number of males in the population is Ns^*, and of outcrossing females is $N(1-s^*)(1-k)$. Hence the expected number of matings to a male by outcrossing is $(1-s^*)(1-k)/s^*$.

GM has rs sons and $r(1-s)$ daughters.

Each daughter has r offspring.

Each son has $rk(1-s)/s$ offspring by his sisters, and $r(1-s^*)(1-k)/s^*$ offspring by outcrossing.

Since, when an individual has a child, one half of his autosomal genes are transmitted to that child,

$$T = \frac{r^2}{4} \left[(1-s) + \frac{s(1-s)k}{s} + \frac{s(1-s^*)(1-k)}{s^*} \right].$$

If s^* is an ESS, then $(\partial T/\partial s)_* = 0$. This gives

$$s^* = (1-k)/2. \tag{9.4}$$

Note that when $k = 0$ (random mating), $s^* = \frac{1}{2}$, and when $k = 1$ (brother–sister mating), $s^* = 0$. The latter result arises because it has been assumed that there are always enough males in a family to ensure that the females are mated. If we allow for finite family size, and the probability that a family may contain no males, equation (9.4) exaggerates the bias in the sex ratio with inbreeding.

Hamilton (1967) obtained equation (9.4) for a similar model. He supposed that the population consisted of groups of n families whose members mated randomly. A female then has a probability $1/n$ of mating her brother. Hamilton obtained $s^* = (n-1)/2n$, which becomes as equation (9.4) if we put $k = 1/n$.

Thus even partial inbreeding produces a large distortion of the evolutionarily stable sex ratio. A similar distortion exists if inbreeding occurs only in some generations; for example, a population which mates both randomly and brother–sister in alternate generations has a value of $s^* = \frac{5}{16}$ (an unpublished analytical result).

F The sex ratio when the value of a male or female varies with circumstances

Fisher allowed for a differential cost of males and females, but not for the possibility that there might be circumstances in which it would pay to produce one or other sex. Trivers & Willard (1973) pointed out that if there is a greater variance in fitness for one sex than for the other, and if high parental investment puts the offspring at the top rather than the bottom end of the fitness distribution, then a parent who is able to make a high investment will benefit by producing offspring of the high-variance sex, and a parent who can only make a low investment should produce offspring of the low-variance sex.

As an application of this general idea, they considered polygynous mammalian species, in which the variance of reproductive success may be greater for males than for females. They, therefore, predict that females in good reproductive condition should produce sons, and in less good condition daughters. Their paper quotes data on deer, pigs, sheep, dogs, seals, and humans, showing that the sex ratio is biased in the expected direction.

Charnov (1977*a*), quoting earlier work by Chewyreuv (1913), Assem (1971) and others, uses an argument similar to that of Trivers & Willard (1973) to explain the behaviour of solitary hymenopteran parasites. It is a common observation that such parasites produce sons in small hosts and daughters in large hosts. Chewyreuv found that it was relative and not absolute size that was important. An ichneumonid offered pupae of *Sphinx ligustri* (large) and *Pieris brassicae* (medium) produced females from the former and males from the latter. When offered pupae of *Pieris* alternated with *Vanessa levana* (small), the same ichneumonids produced females from *Pieris* and males from *Vanessa*. Charnov (1978*a*) shows that this behaviour is optimal if the *relative* advantage to males of being raised in a large host is less than the relative advantage to females. As in Trivers & Willard, the relevant feature is which sex has the greater variance of fitness.

The third example to be discussed is different in kind. Freeman, Klikoff & Harper (1976) studied the sex ratio in natural populations of dioecious plants: salt grass, *Distichlis spicata*; meadow rue, *Thalictrum fendleri*; box elder, *Acer negundo*; mormontea, *Ephedra viridus*; and shadscale, *Atriplex confertifolia*. In all five species they found an excess of males in the drier and more exposed sites, and of females in the wetter sites. This makes ecological sense, in that female fecundity in these plants depends on a prolonged growing period possible only in the wetter sites, whereas pollen dispersal would be more effective in drier sites. The differences are large; for *Thalictrum*, the sex ratio varied from 6.57 in dry sites to 0.24 in moist and shady ones. What is not clear is the mechanism; there may be differential mortality, or sex in these plants may be environmentally determined. However, in *T. fendleri* sex is genetically determined, the males being XY and the females XX (Westergaard, 1958).

Evolution of a system of this kind by differential mortality is hard to understand. Progeny survival is likely to be under the control of genes in the progeny rather than in the parent. If so, how could genes favouring the death of males in moist sites, or of females in dry ones, be established? A plausible answer is that seedlings tend to be closely related to their neighbours. If sex differentials in fitness are high enough, a gene causing a male in a moist site to sacrifice itself for a sister could spread by kin selection.

However, environmental sex determination seems a more natural explanation. This raises the question of why environmental sex determination is not commoner than it is. Charnov & Bull (1977) have argued that it is to be expected whenever unpredictable features of the environment dictate to the individual that it would pay to be either a male or a female. The most obvious circumstance of this kind is the local sex ratio where an individual settles. It is possible that environmental sex determination may be commoner among sessile organisms than has been suspected.

It would also pay a parent to produce offspring of whichever sex is locally rarer. There is some evidence that this can happen. Snyder (1976) reports some remarkable observations on the woodchuck, *Marmota monax*. The sex ratio among young adults in natural populations was close to 1:1. In one population, Snyder removed approximately half the breeding females; the sex ratio among the young adults in the next year was forty males to eighty-nine females.

G   Parent–gamete competition: meiotic drive

Some of the most striking distortions of the sex ratio arise because genes on the X or Y chromosomes are expressed in the gametes or, more usually, during meiosis. A gene on a sex-determining chromosome making that chromosome more likely to participate in fertilisation will increase in frequency until checked by some counterbalancing force. A particularly illuminating case is the Y-linked gene M^D in the mosquito *Aedes aegypti* (Hickey & Craig, 1966; Wood & Newton, 1976). Males carrying the gene produce an excess of sons because of X chromosome breakage during male meiosis. This breakage causes a reduction in sperm number, but not the production of aneuploid sperm, mainly because spermatids carrying deficient X chromosomes degenerate.

If this were the whole story, natural populations of *A. aegypti* would consist mainly of males, and would be on the verge of extinction. However, there are also resistant X chromosomes able to suppress the action of M^D; in fact, eight categories of X chromosomes have been identified, varying from highly sensitive to fully resistant. As one would expect, X chromosomes from Africa and central America, where M^D is commonly found, tend to be resistant.

Clearly, in the presence of a gene such as M^D, there will be strong selection favouring genes on the X chromosomes which suppress M^D; there will also be selection for suppressor genes on the autosomes, because, as the population sex ratio is distorted in favour of males, the production of daughters will be favoured. Hamilton (1967) suggested that the Y chromosome is commonly inert in animals because there has in the past been selection to suppress Y-linked driving genes.

Driving genes on sex-determining chromosomes could also cause the evolution of a change in the sex-determining mechanism. For example, if a driving Y chromosome became common, there would be selection favouring the conversion of XY individuals into females. If chromosomal differentiation between X and Y were not too great, this would lead to another pair of chromosomes taking over the sex-determining role. In fact, if it were not for the conflict between genes in the parent and in the gamete, exemplified by meiotic drive and its suppression, it is difficult to see why sex-determining mechanisms should ever change, as they certainly do (see Wagoner, McDonald & Childress, 1974, for an example of two different mechanisms in a single species).

H   Parent–offspring competition

Trivers (1974) has pointed out that if the sex ratio is distorted from
$1:1$, for example by an excess of females, a gene in a female
converting that female to a male would be favoured by selection,
because whichever is the rarer sex (in this case the male) has a higher
expected reproductive success. At first sight one might suppose that if
sex ratio were determined by genes in the offspring, the sex ratio
would be $1:1$, because it always pays to be the rarer sex. However, as
Trivers points out, an offspring which insists on being a male reduces
the number of other offspring its mother can produce. If males cost x
times as much as females, the mother's (or father's) preferred sex
ratio is one male to x females. So long as the population sex ratio
remains $1:x$, a gene converting females to males (if $x > 1$) is favoured.
If the offspring displaced are full sibs, then the break-even point for
genes acting in the offspring turns out to be a sex ratio of $1:\sqrt{x}$.

The idea is ingenious, and the argument seems to be correct. It is
not clear whether the selective force proposed has had significant
consequences. A related idea concerning the sex ratio in eusocial
hymenopterans has been suggested by Trivers & Hare (1976). Here
the conflict is between the queen and her worker daughters. Genes
acting in the queen are favoured if they cause equal investment in
reproductive sons and daughters. Since males are smaller than
females, this implies a numerical excess of sons. (Of course, genes in
the queen will also be selected if they cause the queen to divide her
own energies between reproductive and non-reproductive (worker)
offspring in such a way as to maximise the final number of
reproductives produced. But this allocation problem is not the one
that Trivers & Hare were concerned with.)

In contrast, genes in workers will be favoured if they cause the
workers to invest more in their reproductive sisters than in their
brothers. This is because a worker has $\frac{3}{4}$ of her genes in common with
her sisters, and only $\frac{1}{4}$ in common with her brothers – a point which
was at the heart of Hamilton's (1964) interpretation of the evolution
of social hymenopterans. Using Hamilton's concept of inclusive
fitness, Trivers & Hare argue that the workers will prefer to invest

three times as much energy in sisters than in brothers, whereas the queen prefers a 1:1 investment ratio.

Trivers & Hare consider two complications. First, it pays a worker better to raise her own son, or even a nephew (i.e. the son of another worker) rather than her brothers, and in many eusocial hymenopterans there are laying workers producing sons. Secondly, the original female may have been multiply inseminated, in which case the coefficients of relationship change. They find the preferred sex ratio allowing for these factors; Charnov (1978b) has reanalysed the problem, using a more orthodox population genetics approach, and gets results which are closely similar but not identical.

It would seem, then, that if the actual sex ratio approximates to 1:1, then the ratio is being controlled by the queen; if to 1:3, then control is by the workers. Trivers & Hare present evidence, using biomass rather than numbers so as to measure investment, that the actual ratio in eusocial hymenopterans is never 1:1, and often approximates to 1:3. Unfortunately, the interpretation of biological data is rarely unambiguous. Alexander & Sherman (1977) have pointed out that the observed bias in the sex ratio could arise from another cause. If there is competition between brothers for matings, then, as explained above, we would expect a sex ratio biased in favour of females even if the queen is in control. For the time being the matter must be left open.

Appendix: The evolution of stable sex ratios

Suppose that a pair can produce m male and f female offspring, where (m, f) is constrained to lie within a fitness set. We seek a sex ratio (m^*, f^*) which is evolutionarily stable.

Consider a random-mating population in which typical pairs produce m^* sons and f^* daughters. A rare dominant gene M causes females to produce m sons and f daughters, and is not expressed in males (an exactly similar conclusion follows for a gene expressed only in males). The frequency of $M/+$ females is P and of $M/+$ males is p; P and p are small, and since the population is random-mating we can ignore M/M individuals. Continuing to ignore terms of the order P^2, we have:

Matings		Frequency	Total offspring			
			male		female	
♀	♂		$M/+$	$+/+$	$M/+$	$+/+$
$M/+$	$+/+$	P	$\frac{1}{2}mP$	$\frac{1}{2}mP$	$\frac{1}{2}fP$	$\frac{1}{2}fP$
$+/+$	$M/+$	p	$\frac{1}{2}m^*p$	$\frac{1}{2}m^*p$	$\frac{1}{2}f^*p$	$\frac{1}{2}f^*p$
$+/+$	$+/+$	$(1-P-p)$	—	$m^*(1-P-p)$	—	$f^*(1-P-p)$
		Total	$m^* + P(m-m^*)$		$f^* + P(f-f^*)$	

If the corresponding frequencies in the next generation are P', p', then

$$P' \simeq \frac{1}{2}\frac{m}{m^*}P + \frac{1}{2}p,$$

$$p' \simeq \frac{1}{2}\frac{f}{f^*}P + \frac{1}{2}p.$$

Adding the two equations,

$$P' + p' \simeq P + p + RP,$$

where $R = \frac{1}{2}(m/m^* + f/f^*) - 1$.

If $m = m^*$ and $f = f^*$, then $R = 0$, and the frequency of M in the population does not alter (as is required, since M has no phenotypic effect). If (m^*, f^*) is to be evolutionarily stable, then for any values of (m, f) not equal to (m^*, f^*), R must be negative; the frequency of any mutant M will then decrease. That is, $m/m^* + f/f^*$ must be a maximum when $m = m^*, f = f^*$; as MacArthur (1965) pointed out, this requires that the product $m^* \times f^*$ be a maximum.

Thus in a random-mating diploid population in which the sex ratio is determined by autosomal genes acting in the parents, selection will maximise the product of the numbers of sons and daughters produced by an individual. The same conclusion holds for genes on the X chromosome, or genes in a haplo-diploid population, provided that the genes act in the female parent.

10 *Sexual selection*

Since Darwin (1871) first introduced the idea of sexual selection, authors have differed as to the range of processes they would include under this heading. As soon as anisogamy has evolved, different selective forces may act on males and females; it is these differential forces with which I am concerned in this chapter. Darwin himself distinguished between selection arising because of direct competition between males, and that arising because females may choose one type of male rather than another.

A   The concept of 'female choice'

It is convenient to start with a brief description of courtship behaviour in the fruit fly *Drosophila subobscura* (Maynard Smith, 1956). If a male and a virgin female are brought together in a suitable mating chamber in the light, the male approaches the female and circles round her until he is facing her head to head. The female then steps rapidly to one side or the other, and the male side-steps likewise, so as to keep his position facing her; the female may reverse her direction, so that a rapid side-to-side dance ensues. After the dance has lasted for one or a few seconds, one of two things may happen. The female may stand still, in which case the male circles round behind her and mounts, and mating takes place; or the female may jump or fly away.

The female is much more likely to fly away if the male is inbred, old or injured. In any of these cases, the male is less able to keep up with the rapidly moving female. It is a strong presumption that the female accepts a male who keeps up adequately during the dance, and rejects one who does not. The female is, therefore, extremely discriminating; in contrast, a male will dance with and attempt to mount a blob of wax on the end of a bristle, moved in an appropriate side-to-side manner and then held still.

This seems a clear case of 'female choice'. Before discussing the selective forces causing the female to behave in this way, I must say something about the word 'choice'. When we say that a man 'chooses' to do something, we mean that he could have done otherwise; further, we imply that if different inducements or arguments had been put to him, or if he had been brought up differently, he *would* have done otherwise. By analogy, if I say that a female fruit fly chooses to mate with an outbred rather than an inbred male, I mean that she could have done otherwise; more specifically, I mean that simple genetic changes in a female fruit fly could cause her to lose the discrimination or even to reverse it. In other words, when we speak of human choice, we imply actions which are individually flexible; when we speak of 'choice' in the context of the evolution of behaviour, we imply actions which are flexible on an evolutionary time scale.

Of course, it is quite possible to discuss the evolution of courtship behaviour without using the word 'choice', and it may be best to do so. However, it is *not* possible to discuss rationally the selective forces responsible for the evolution of particular behaviour patterns unless we specify clearly what types of variation in behaviour we consider to be evolutionarily possible. In the context of courtship behaviour, what this means is that we must consider possible changes both in male and in female behaviour.

To bring this point home, consider the following question. Is it reasonable to speak of female choice in the red deer, in which stags fight one another to hold a harem of hinds? I suggest that it is, or, what in my view amounts to the same thing, I suggest that an explanation of red deer behaviour requires not only an explanation of why stags fight, but also why hinds remain in the harem (in fact they do not always do so).

Reverting to *D. subobscura,* why do females behave so as to discriminate between different conspecific males, whereas males hardly discriminate at all? Bateman (1948) suggested the following answer to this question. A male can mate many times in each day. Each conspecific mating increases his fitness. He has little to lose by courting or mating with unsuitable females (e.g. females of a related species). In contrast, a female can obtain in a single mating enough sperm to fertilise all the eggs which she will lay in her lifetime. She cannot increase her fitness by mating more than once (in fact, a mated female *D. subobscura* only mates a second time if the first mating failed to lead to insemination), but can do so by selecting as a mate a male of high fitness. If there is a correlation between the dancing ability of males and their fitness as fathers, it will pay the female to discriminate by dancing.

There are two possible objections to this orthodox answer. First, there may be little additive genetic variance of fitness; secondly, some traits apparently chosen by females seem unlikely to correlate with fitness. I will discuss these points in turn.

There is little additive genetic variance of fitness

There are both theoretical and observational grounds for thinking that there will be a low parent–offspring correlation for fitness (in effect, additive genetic variance for fitness is rapidly depleted; for a discussion of this point, see Falconer, 1960). Therefore, a female who selects as a mate a male of high fitness does not increase the expected fitness of her own offspring. The difficulty is a serious one. If the male contributes anything other than gametes to his children (e.g. protection, feeding, a good territory), then the difficulty disappears. But the choice by females of particular males, or males in particular territories, in a 'lek' situation (in which a group of males compete in a display area for the attention of females) cannot be explained in this way; nor indeed can choice by female *D. subobscura*.

In lek species, it is often not clear whether the female is choosing the male or the territory which he holds. Wiley (1973) has shown that in the ruffed grouse, *Bonasa umbellus,* males holding a central territory in the lek obtain most matings, and the same is probably true of some other lek species. If so, the often somewhat bizarre characteristics of the male may be more important in enabling them to compete successfully with other males for a central territory than in directly attracting females (although there are difficulties in understanding how such inter-male contests are in fact settled). In the present context, however, the important point about lek species is that the female gets nothing from the male except genes. It, therefore, would not pay a female to select a male from a central territory unless there is some additive genetic variance for whatever characteristics enable a male to obtain a central territory.

The only way out of this difficulty I can see is to argue that, although small, the additive heritability of fitness is not zero. Both recurrent harmful mutations and transient polymorphism resulting from the establishment of new favourable mutations will contribute additive fitness variance, and so make female choice worthwhile. It is curious that the same two features of natural populations must be invoked to explain the evolution of recombination (see p. 119).

If most of the genetic variance of fitness is not additive, one might predict that a female should select as a mate a male genetically unlike herself. I know of no evidence for such behaviour, although behaviour patterns with the same consequence were discussed in Chapter 8. The nearest approach is the 'rare male phenomenon' in male Diptera (Ehrman, 1972), in which females select as mates males of the minority type in the local population. For most females, this would have the consequence of ensuring a mating with a genetically unlike male.

The choice of 'handicaps'

A female which chooses a male capable of performing a complex dance or ritual chase is at least likely to choose one of high individual fitness, whatever the doubts may be about the mean fitness of her offspring. But what of a peahen which chooses a peacock with a large and elaborate tail? The possession of such a tail is likely to interfere with flight and render the male more subject to predation. Although there is no direct evidence on this point for peacocks, Darwin (1871) pointed out that if a characteristic is confined to the male, this implies that it is disadvantageous. Selander (1972) gives convincing evidence that in the polygamous great-tailed grackle, *Quiscalus mexicanus,* the proportion of males falls from 50% among nestlings to less than 30% in the adult population. Why should a female grackle choose to mate with a male whose sons, if they resemble their father, will suffer a high mortality before breeding?

Fisher (1930) suggested that a process I shall call 'self-reinforcing choice' can answer this question. Suppose there is a polygamous population in which the majority of females do in fact prefer males with long tails, and consider the fate of a rare mutant type of female which selects males with short tails. The mutant female will have sons with shorter-than-average tails, which are, therefore, more likely to survive to breed. These sons, however, will be of low fitness, because typical females in the population will not select them as mates. Thus the existence of males with excessively long tails, and of females preferring such males, would be stable once it had evolved. (Note that, as in his discussion of the sex ratio, Fisher uses what is in effect an ESS argument, see p. 147.)

This leaves unanswered how the preference arose in the first place. Many answers are possible. One is as follows. Suppose that at some earlier time two sympatric populations were evolving into good species, and that hybrids between them were of low fitness. If these populations differed in tail length, then females in the long-tailed population would be favoured if they chose as mates the males with the longest tails. This could soon become self-reinforcing by Fisher's mechanism, even in the absence of the other population, and could then lead to a continued exaggeration in tail length beyond the range initially present.

It is worth pointing out that there is again some doubt as to whether there is sufficient additive genetic variance to make this process work. If there is strong directional selection, because of female choice, in favour of longer tails, most of the additive variance for the character will be fixed, and it will no longer pay the females to select males with long tails. Apart from this difficulty, there is no reason to doubt Fisher's argument, which has been developed mathematically by O'Donald (1962).

The same difficulty arises in some fish. Haas (1976) studied the 'annual' cyprinodont fish *Notobranchius guentheri* in ephemeral bodies of water in West Africa. The females are uniform brown, and the males more brightly coloured, with a vivid carmine tail. Females pair with a male only long enough for a batch of eggs to be laid and fertilised. The males are not territorial, but a dominance hierarchy exists, with the more dominant males interfering aggressively with the reproduction of other males. There is virtually no courtship; the female approaches a male and mating takes place almost at once.

Haas showed that in experimental conditions females approached males more readily than they approached other females, and that males were more taken by a predator (a heron) than females; both effects depended in part on the fact that a fish with a bright red tail is more easily visible in turbid water.

In a sense there is nothing very surprising about these results. But the question remains. Why do not males lose their red tails and so become less vulnerable to predators? Why do not females seek out males without red tails, and so have sons which escape predation? The case is not unique. Haskins *et al.* (1961) showed that female guppies choose the most brightly coloured males, which are most susceptible to predation. Fisher's idea of self-reinforcing choice seems the best way to explain these facts.

However, Zahavi (1975) has offered an alternative explanation of sexual characteristics which have been exaggerated to the point of being a serious handicap. He argues as follows (I use anthropomorphic terms, but Zahavi's argument can be formulated without them, albeit rather clumsily). A peahen wishes to select as the father of her children a male with genes for high fitness. Unfortunately, there is no way in which such genes can be directly recognised. However, a male who carries a handicap, for example a long and unwieldy tail, would not have survived to breed unless he was of high fitness in other respects. Therefore, by choosing a male with a long tail the female ensures that she at the same time chooses a male with genes for high fitness.

Zahavi calls this the 'handicap principle'. It has been criticised by Davis & O'Donald (1976) , and by Maynard Smith (1976*d*). The essential difficulty is as follows. A female who selects a male with a handicap ensures that her children will inherit genes for high fitness from their father, but she also ensures that they will inherit genes for the handicap. On balance, will they be worse off or better? This does not seem to me to be a question which can be settled by verbal argument. One must set up a genetic model, and see what happens.

There are of course difficulties. The main one is this: suppose that genes for a handicap, and for choosing males with a handicap, increase in frequency, is the increase because of Zahavi's handicap principle, or of Fisher's self-reinforcing choice? This difficulty can be met by seeing whether handicap and choosing genes increase in frequency *when rare*; if they do, then it is because of Zahavi's principle, because Fisher's process only works when choosing is already common. I adopted this approach, in a simple model with separate loci determining general fitness, handicap and female choice (Maynard Smith, 1976*d*). To make it more likely that the handicap principle would work, I assumed that the handicap genes were expressed only in males; thus a choosing female has daughters of high fitness with no handicap.

I was unable to find any plausible situations in which Zahavi's principle would operate. Genes for the handicap, or for choosing males with a handicap, always decreased in frequency when rare; when common, such genes could often increase in frequency further, as predicted by Fisher's argument. The two situations in which the handicap principle would operate are

(i) genes for the handicap non-additive; genes for fitness additive, and
(ii) handicap non-genetic.

I do not regard either of these as plausible. In reply, Zahavi (1977) has suggested a further verbal model, of which he says 'it is obvious that this model can operate'. Since his new model assumes non-additive inheritance of fitness, it seems to me obvious that it would not. I still think it possible that in some modified form the handicap principle may be important, but I see little point in further discussion until it has been shown to work in at least one plausible genetic model.

I cannot leave the topic of female choice without commenting on Trivers's (1976) suggestion that it may help to account for the maintenance of sexual reproduction itself. He points out that if female choice is so discriminating that 'those males actually mating in each generation have genes which are about twice as good at producing successful daughters as are the genes of the mothers themselves', this would explain why asexual mutants do not spread. This amounts to suggesting that the mean fitness of the offspring of a female which chooses a male may be twice that of an identical replica of that female. It is hard to see how this could be so. There may be some additive genetic variance of fitness, but surely not as much that.

B Parental care

The argument, based on the results from *Drosophila subobscura*, that it is selectively advantageous for males to be promiscuous and females to be selective assumes that a male contributes nothing to his offspring except gametes. As soon as parents guard or feed their young, the argument becomes more complex. Recent interest in the problem has been greatly stimulated by Trivers (1972). He analyses the problem in terms of 'parental investment', defined as 'any investment by the parent in an individual offspring that increases the offspring's chance of surviving (and hence reproductive success) at the cost of the parent's ability to invest in other offspring'.

Trivers's approach has recently been criticised by Dawkins & Carlisle (1976), on the grounds that an animal's behaviour should be determined by the expected pay-off in the future rather than by whether some past investment will be wasted. Trivers could defend himself by saying that he *defined* investment in terms of loss of future reproduction, so that his analysis is in effect prospective. However, I think that the concept of investment can be misleading and is always hard to apply correctly. I have suggested (Maynard Smith, 1977*a*) an alternative way of formulating the problem, in terms of game theory and 'ESSs'. As always, this approach cannot tell us how existing behaviour patterns originated, but only how they are maintained. One main conclusion of the analysis, namely that the sex which contributes most to producing the young may contribute least to caring for them, seems to run contrary to conclusions based on the concept of investment.

My basic approach is to seek for a pair of 'strategies', I_m and I_f, for the male and female respectively, which are evolutionarily stable in the sense that

(i) if most males adopt I_m, it would not pay a female to do other than I_f, and

(ii) if most females adopt I_f, it would not pay a male to do other than I_m.

One point should be made at the outset. I treat 'strategies' as if they are pre-programmed at the start of breeding, and not conditional on the behaviour of the particular partner. For example, I_m might be 'desert the female as soon as the eggs are laid, and attempt to mate with a second female', but not 'desert the female if she is sitting on the eggs, but stay and incubate if she deserts'. There is nothing in the game theory approach which prevents the analysis of strategies of the latter kind (in fact, Maynard Smith & Price (1973) did so in their analysis of contest strategies), but in the present context programmed strategies seem more appropriate, at least for a first attack on the problem.

Consider a species with discrete breeding seasons:

Let P_0, P_1, and P_2 be the probabilities of survival of eggs guarded by 0, 1 and 2 parents respectively.

Let p be the probability that a male which deserts after mating will be able to mate with a second female.

Let V be the number of eggs laid by a female who deserts after laying, and v by a female who guards the eggs after laying. If strategies are pre-programmed, and if guarding requires that the parent start with adequate food reserves, V may be substantially larger than v.

P_0, P_1, P_2, p, V, and v will be treated as constant constraints, subject to which behavioural strategies evolve. For simplicity, I consider only two strategies for each sex, namely $G =$ 'guard' and $D =$ 'desert'. The model has been deliberately kept simple. Some of the complications (e.g. that a male may not be the father of the young he is guarding; that a male may be able both to guard and to mate a second female; that it may pay a female to remate) are discussed later.

The first step in the game theory analysis is to write down the 'pay-off matrix'; that is, the expected number of surviving offspring produced by each sex, according to the strategies adopted. The matrix is shown in Table 7.

I have assumed that a deserting male has only one chance to remate. If a male can mate many times, $V(1+p)$ would become $V(1+p+p^2+ \ldots)=V/(1-p)$. But at this level of analysis the difference is not important.

There are four possible pairs of strategies, and it is a simple matter to write down the conditions which must be satisfied if each is to be an ESS:

ESS 1.   Both G; 'Gibbon'.
$$vP_2 > \quad VP_1, \text{ or female deserts,}$$
and $\qquad P_2 > \quad P_1(1+p), \text{ or male deserts.}$

ESS 2.   Male G, female D; 'Stickleback'.
$$VP_1 > \quad vP_2, \text{ or female guards,}$$
and $\qquad P_1 > \quad P_0(1+P), \text{ or male deserts.}$

ESS 3.   Male D, female G; 'Duck'.
$$vP_1 > \quad VP_0, \text{ or female deserts,}$$
and $\quad P_1(1+p) > \quad P_2, \text{ or male guards.}$

ESS 4.   Both D; 'Fruit fly'.
$$VP_0 > \quad vP_1, \text{ or female guards,}$$
and $\quad P_0(1+p) > \quad P_1, \text{ or male guards.}$

If $P_2 \gg P_1$ (two parents much better than one) and p is not too large, then the ESS will be for both parents to guard. If P_0 is equal to or not greatly less than P_1, the ESS will be for both parents to desert.

I am particularly interested in what I have called the Stickleback and the Duck strategies (although, as will appear below, neither is a wholly appropriate name). The Stickleback ESS is favoured if $V \gg v$ (a female who guards cannot afford to lay as many eggs); if one parent is almost as good as two, and much better than no parent; and

Table 7. *Pay-offs to parents adopting the 'guarding', G, and 'deserting', D, strategies*

		Female	
		G	D
Male	G	vP_2 / vP_2	VP_1 / VP_1
	D	vP_1 / $vP_1(1+p)$	VP_0 / $VP_0(1+p)$

if p is not too large. The Duck ESS is favoured by exactly the same constraints on the effectiveness of parental care, and also if V is not greatly different from v, and p is large. In a sentence, if males have a good chance to remate and if egg laying is not too exhausting, this favours the Duck ESS, and the reverse conditions favour the Stickleback ESS.

It is important, however, to notice that for a given set of values of the constraints both Duck and Stickleback ESSs may exist. Of course, no population can be in both states at once, but a population could evolve to either ESS, depending on the initial conditions. This is unfortunate, because it means that the existing state of a population may depend on initial conditions in an ancestral population no longer available for study. Something can, however, be said. Suppose first that uniparental care has evolved from no parental care. There are three situations to consider:

(i) Fertilisation internal. In this case, the male is unlikely to be present when the eggs are laid, so it is likely to be the female which evolves the guarding habit. There is a second consideration (Trivers, 1972); the male cannot be certain of paternity, so that he has less to gain by guarding than the female. I am not certain that this argument is valid. Thus what we have to compare are not pay-offs to the male and female, but the pay-off to a male who guards and one who deserts. The gain from desertion is the chance of additional mating. But, if paternity is uncertain, the value of additional mating is reduced by exactly the same factor as the value of guarding. It may pay a male to desert if this greatly increases his chance of remating, but confidence of paternity has little to do with it.

Whether because of confidence of paternity, or because the male is absent when eggs are laid, it certainly seems to be the case that in groups with internal fertilisation, male parental care is rare. For example, it is very rare in reptiles, and unusual in fish with internal fertilisation.

It has been argued that lack of certainty of paternity does not of itself select against paternal care. However, there is strong selection on males to increase their confidence in paternity (Parker, 1974). For example, male insects may guard a female with which they have copulated, until the eggs are laid. This has sometimes given rise to male parental care of eggs and young; unfortunately, this could be taken as supporting either argument, because post-copulatory guarding will both ensure that the male is present when the eggs are laid, and increase his probability of paternity.

(ii) Fertilisation external, with pair formation. In this case (e.g. most anurans), both suggested barriers to the evolution of male care are greatly reduced. If, as is likely to be the case, egg laying exhausts the female, the most likely ESS to evolve is the Stickleback ESS. Some evidence that this is in fact so is discussed below.

(iii) Collective external fertilisation. No parental care is likely to evolve, if only because no individual, of either sex, could identify their own offspring.

The model will now be applied to some of the breeding patterns found among vertebrates. In mammals, breeding systems are dominated by female lactation. Males usually cannot feed their offspring. They may, in some instances, contribute to their survival by protection against predators, but in most groups selection on the males has been to increase the number of matings through various systems of polygamy. Monogamy is found in many carnivores, and in some primates. In carnivores, the male can contribute to the survival of his offspring by bringing food to the nursing female or by feeding the young directly. In primates, monogamy is found in those species (e.g. gibbons, marmosets) in which the male can defend a territory large enough to support one female and her young, but not more (Clutton-Brock & Harvey, 1977). The males, in these species, contribute by caring for and protecting their offspring, and by defending a territory, but it is odd that in no case has male lactation evolved.

If mammals are dominated by lactation, birds are dominated by the fact that, in most species, breeding success is not limited by the number of eggs a female can lay (Lack, 1968), but by the number that can be incubated, and the number of young that the parents can feed and protect. It follows that, in terms of the model, V and v are not different, because egg laying does not significantly interfere with the ability of the female to incubate, protect, and feed her offspring. In those species in which the parents feed the young, either in the nest as do most passerines and raptors, or after leaving the nest as, for example, do grebes and divers, much the commonest pattern is for both parents to contribute more or less equally (as predicted by the model, since $P_2 \gg P_1$). If the young are not fed by their parents, it is common for the female only to guard the young. This is to be expected, since $V \simeq v$ and $P_2 \simeq P_1$. Yet there are wide variations in breeding strategy, and only in some cases has polygyny evolved.

Verner & Willson (1966) and Orians (1969) proposed that polygyny will evolve if the qualities of the territories held by different males vary so greatly that it is better for a female to mate with an already mated male in a good territory than with an unmated male in a less good territory. There is support (reviewed by Selander, 1972) for this theory for species in which the males hold territories within which one or more females nest. First, in three such species (the long-billed marsh wren, the red-winged blackbird, the great-tailed grackle), it is known that females do often choose to breed with already mated males when unmated males are available. Secondly, there is evidence for the red-winged blackbird (Haigh, quoted by Selander, 1972) that the number of young actually fledged per female is higher for polygynously than monogamously mated females.

Selander suggests that this pattern of breeding evolves in species in which suitable nesting sites rather than food are limiting; if suitable sites are very patchily distributed, there might well be a large variance between the values of territories held by different males, as required by the Verner–Willson–Orians theory.

However, the theory does not help us to explain the type of polygyny found, for example, in some grouse (Tetraonidae) in which the male contributes nothing other than a mating. Breeding systems in the grouse have recently been reviewed by Wiley (1974). Some species form more or less permanent pair bonds, although only in the red grouse, *Lagopus lagopus,* does the male contribute significantly to the care of the young. Others are polygynous, the males either congregating to display in a lek, as in the black grouse and ruffed grouse, or (in forest species) displaying from widely dispersed sites. It is far from clear why these differences have evolved.

In the waterfowl (Anatidae), reviewed by Kear (1970), it seems clear that uniparental care, as found in the river ducks (*Anas*), diving ducks (*Aythya*), and stifftails (*Oxyura*) has evolved from biparental care found in the larger geese and swans (*Anser, Cygnus*) in which the male stays on guard near the nest and protects the young. The obvious explanation for the change is that the larger species are able actively to drive away most potential predators, whereas the smaller species protect their young by nest concealment and feigning injury; in the latter case, one parent is as good as two in guarding the young. In most of the smaller species, the male deserts soon after mating. However, the advantage gained is probably not an increased chance of a second mating (as implied in the model) but an increased probability of survival (Bellrose *et al.*, 1961).

Most of the rare cases of polyandry among birds occur among the waders (Jenni, 1974). The American jacana, *Jacana spinosa* (Jenni & Collier, 1972), breeds throughout the year. The female holds and defends a large territory, within which several males hold separate territories. The males construct the nest and incubate and care for the young. The female is 75% heavier than the male. Jenni (1974) suggests that this unusual breeding strategy may have evolved from the habit of double-clutching found in some waders. In the sanderling (*Calidris alba*) and Temminck's stint (*C. temminckii*) it is common for the female to lay one clutch of four eggs which is incubated and reared by the male, and then to pair with a second male and lay a second clutch which she incubates and rears herself. This can be seen as an adaptation to a short Arctic breeding season with an abundant food supply.

With double clutching, it is clear that the male must care for the first clutch. It is then a short step to the situation found in the spotted sandpiper, *Actitis macularis,* in which, if there is a local excess of males, a female will produce a series of clutches cared for by different males, staying with the last male to help rearing. Once this stage is reached a female can increase her fitness by driving off other females in order to acquire additional males to rear her eggs, leading to the situation in the jacana (although it must be remembered that double clutching is found in Arctic breeding waders, whereas the jacana breeds continuously in a tropical environment).

I argued earlier that the reason why the Duck ESS is much

commoner in birds than the Stickleback ESS is that the capacity of the female to lay eggs is not limiting, so that $V \simeq v$. To conclude this brief review of breeding strategies in birds, consider two species in which the female capacity to lay eggs is probably limiting, and in which the male cares for the eggs. *Rhea americana* (Bruning, 1973) has a defined breeding season, at the start of which the males fight to establish a dominance hierarchy. The dominant males collect harems of up to fifteen females, copulating with each female every two to three days. The females lay eggs in a nest constructed by the male, until up to ten days have passed and up to fifty eggs have been laid. The male then drives away the females, and commences to incubate the eggs. The females will then join another male, and may in fact consort with and lay eggs for a number of males in series during the season.

Bruning showed that eggs laid early in the season have the best chance of survival. It is, therefore, easy to see why the strategies described are evolutionarily stable, in that neither sex could benefit by behaving differently. A dominant male gets as many eggs as he can raise, probably fathered by himself, and laid as early in the season as possible; a female ensures that as soon as she can lay another egg, a male will start incubating it.

In the mallee-fowl, *Leipoa ocellata* (Frith, 1962), the male constructs a large mound of sand and dead leaves as an incubator for the eggs. The female lays eggs over a period of three to four months, and may lay up to thirty eggs, each 10% of her own weight. During the laying period the male alone constructs and tends the mound. The pair are monogamous for life. Although this is a very different breeding strategy from that of the rhea, in both cases the female's capacity to lay eggs is limiting, and the male cares for the young.

Of the cold-blooded vertebrates, the anurans and the bony fishes are of particular interest. In the frogs, it seems to be about equally frequent for the male and for the female to care for the young. The group is a particularly favourable one for investigating why it is sometimes the Duck ESS and sometimes the Stickleback ESS which has evolved. The bony fishes have been excellently reviewed by Breder & Rosen (1966), of whose data Table 8 is a partial summary. By and large, the figures bear out the prediction that with external fertilisation the Stickleback ESS should be favoured. However, one important modification of the model is needed before it adequately describes the situation in many of the families in the first column of the table. A male who, after mating, cares for the eggs does not necessarily reduce his chances of mating with a second female. For example, in the stickleback itself, a male may attract a succession of females to lay eggs in his nest. For many of the families listed, it is not clear whether the male is typically caring for the offspring of one female or of many.

Table 8. *Numbers of families of bony fishes in which (in some or all species) one or other parent cares for the young (data from Breder & Rosen, 1966)*

Fertilisation external			Fertilisation internal			No parental care
♂	♀	both	♂	♀	both	
28	6	8	2	10	0	191

C   Monogamy, polygamy and sexual dimorphism

There is an obvious difficulty in arguing that marked sexual dimorphism has been produced by female choice in monogamous species with a 1 : 1 sex ratio. Since every male will obtain a mate, what is to be gained by extreme epigamic characteristics? Darwin (1871) was aware of this difficulty. He suggested that a male which is preferred by females will start breeding earlier in the season, and this is likely to increase his fitness, both because an early start is in itself advantageous and because the first females to be ready to breed are likely to be the best mates. Similarly, it will pay a female to be selective if by so doing she obtains a better-than-average mate.

It is hard to evaluate Darwin's argument. In its favour is the fact that it need not rely on the presence of additive genetic variance of fitness. In monogamous species it is common for both parents to contribute to the care of the young. Therefore an individual choosing a fit mate is choosing a good producer for his or her children. There is, however, an alternative selective force which may be responsible for epigamic characteristics; this is the advantage of establishing a pre-fertilisation barrier to hybridisation during the early stages of speciation (Dobzhansky, 1937). It is likely that in a group such as the ducks, in which the males contribute little in the way of parental care, this advantage has been more important than Darwinian sexual selection.

The most convincing evidence in favour of Darwinian sexual selection in a monogamous species comes from the work of O'Donald (1976 and earlier) on the Arctic skua in the Shetlands. The species is polymorphic, with dark, intermediate and pale forms of both sexes. The species is monogamous, and a pair which has bred together in one year usually stays together subsequently, although some 15% of birds breeding together for the first time change mates in the next breeding season. Females breeding for the first time prefer dark males, possibly because they hold larger territories. Consequently, in their first season dark males breed earlier than pale ones, and are substantially more successful in raising young. Thus the situation in the Shetlands fits Darwin's proposed mechanism very closely. It is not clear what counterbalancing advantage of the pale form maintains this polymorphism; there is, however, an increase in the frequency of the pale form as one moves northward, so the polymorphism may be maintained by spatial variation in selection.

The intensity of sexual selection, either through female choice or male competition, is far greater in polygamous species. One consequence of this is that the degree of dimorphism in size is in general much greater in polygamous than in monogamous species (for a recent review of the data on primates, see Clutton-Brock & Harvey (1977); on grouse, see Wiley (1974); for birds generally, see Verner & Willson (1969); for evidence that in the lizard *Anolis garmani* larger males do get more copulations, see Trivers (1976); for reversed size dimorphism in polyandrous birds, see Jenni (1974)).

Although, at least in vertebrates, Darwinian sexual selection seems to be the most important force producing sexual dimorphism, and to be responsible for the most extreme examples, it is not the only one. Another well-understood process is ecological differentiation, usually in feeding, between the sexes. It is important to understand that this does not require a group selection explanation in terms of maximising the range of food available to the species. Consider, for example, the typical case of differentiation in the size of the food taken. In any situation of 'ecological release', the food size typically taken by a species will be in shortest supply, and individuals taking either larger or smaller food will be favoured by selection. If there is, initially, some size difference between the sexes, with, for example, males larger than females, then small females will be fitter on average than large ones, and large males than small ones. Selection, therefore, will exaggerate any pre-existing size difference.

Selander (1972) has given an excellent review of this process in birds. He points out that it is often difficult to tell whether Darwinian sexual selection or ecological differentiation is responsible for a particular difference. In cases such as the huia, *Neomorpha acutirostris,* and some woodpeckers, only trophic structures differ between the sexes. The causal connection with ecological release is confirmed by the fact that, in melanerpine woodpeckers, the degree of dimorphism in bill size is greater in insular than in continental species. In these cases, ecological differentiation is clearly the cause of dimorphism; in contrast, the house sparrow, *Passer domesticus,* is monomorphic in bill size and skull dimensions, but dimorphic in the size of other body parts.

An interesting but unexplained fact (Amadon, 1959) is that in four separately evolved predatory groups (hawks, owls, frigates, and skuas) the females are larger than the males. This cannot be interpreted merely as an exaggeration of a pre-existing difference; for example, the skuas are descended from gulls, in which the male is usually slightly larger than the female. The large female size may be an adaptation to laying a large egg but, if so, the reason for this association between large females and the predatory habit is not obvious.

Darwinian sexual selection and ecological differentiation do not exhaust the selective forces responsible for sexual dimorphism in size. This is brought out by Ralls (1976), who discusses those mammals in which the females are larger than the males. There are no known polyandrous mammals, so competition for mates is unlikely to be important in these cases, and it is hard to account for all the known cases in terms of any one selective mechanism.

One last feature of size dimorphism must be mentioned, although it is hard to explain. Rensch (1960) points out that there is often a correlation between the absolute size (as measured, e.g., by adult female size) and sexual dimorphism in size (i.e. male:female size). A recent example is shown in Figure 21.

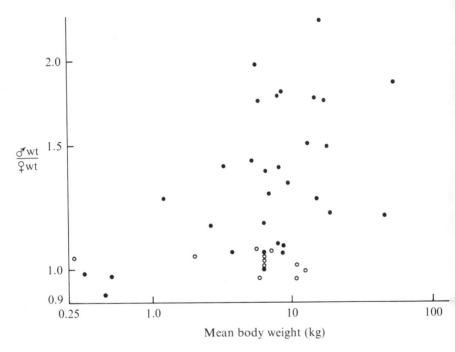

Figure 21. Sexual dimorphism in primates, as a function of body size. Open circles, monogamous species; closed circles, polygamous species. (After Clutton-Brock & Harvey, 1977.)

Why should this be? There is a simple but not wholly satisfactory explanation. In any polygynous species, there will be intense directional selection for increased size of males; selection on females will be normalising, and usually less intense. Since most genes affecting size are not sex limited, the effect of such selection will be to increase both male and female size; since some genes are sex limited, size dimorphism will also increase. Thus polygyny will result in an increase both of absolute size and of dimorphism. Although this pattern of selection may contribute to the correlation by producing large, dimorphic polygynous species and smaller, less dimorphic monogamous ones, there is probably more to it than this.

11 *Mutation*

A   Are mutation rates optimal or minimal?

Two views have been taken about the level of mutation. One is that the actual levels observed are in some way a compromise between selection for reduced levels because of the deleterious effects of most mutations on individual fitness, and selection for increased levels because of the evolutionary value of mutations. The other is that actual levels are as low as is physiologically practicable without excessive cost in energy and time spent replicating.

The first quantitative analysis of the problem was by Kimura (1960, 1967). In the later paper he argues that there will be intra-group selection for reduced mutation rates because most mutations are harmful, and inter-group selection for increased rates because evolution cannot proceed without new mutations. He seeks the optimal compromise by asking what value would minimise the total genetic load L, taken as the sum of the mutational and substitutional loads, and concludes that the rate will be optimal when $M = E$, where M is the mutation rate per gamete per generation, and E the rate of allelic substitution per generation. However, Kimura does not offer any reason why selection should minimise L; I shall return to this question later.

Kimura also discusses the second possibility – that mutation rates are minimal. He points out that observed spontaneous rates differ greatly between organisms, which argues against the view that they are at a physiological minimum, at least in those with the higher rates. He also suggests that, if any mutants are found with rates lower than that typical for the species, they will have deleterious pleiotropic effects. Since he wrote, modifiers lowering spontaneous mutation rates have been found in micro-organisms (references in Drake, 1974); it is still not clear whether they necessarily have deleterious effects on replication rate, although they are known to do so in some cases.

Leigh (1970) argued that in a sexual population the only effective selection would be that tending to minimise the rate, a conclusion he based on the general ineffectiveness of inter-group selection. However, he pointed out that in an asexual population selection would optimise the rate, in Kimura's sense of making $M = E$. I give a modified version of Leigh's argument in the appendix to this chapter. It holds only for an infinite population whose rate of evolution is imposed by the rate of environmental change (i.e. by the rate at which new alleles become selectively advantageous). It may be that the optimal mutation rate in an infinite population is the one most likely to be encountered in a finite one, although I do not see any easy way to prove this.

It is a clear prediction of Leigh's argument that spontaneous mutation rates should be higher in asexual species than in related sexual ones (provided, of course, that the different systems of reproduction have persisted for long enough to permit an evolutionary adjustment of the mutation rate). Unfortunately, I know of no evidence bearing on this prediction.

Spontaneous mutation rates are not easy to measure. Some of the more reliable estimates are given in Table 9. The most striking features are the constancy of the rate per genome for microorganisms, and the much higher value for *Drosophila*. It is therefore natural to ask whether the two sets of values are really comparable. The value for *Drosophila* (Mukai *et al.*, 1972) is based on deleterious

Table 9. *Comparative spontaneous mutation rates (from Drake, 1974)*

Organism	Base pairs per genome	Mutation rate per base pair replication	Mutation rate per genome per generation
Bacteriophage Lambda	4.7×10^4	2.4×10^{-8}	0.001
Bacteriophage T4	1.8×10^5	1.1×10^{-8}	0.002
Salmonella typhimurium	3.8×10^6	2.0×10^{-10}	0.001
Escherichia coli	3.8×10^6	4.0×10^{-10}	0.002
Neurospora crassa	4.5×10^7	6.2×10^{-11}	0.001
Drosophila melanogaster	4.0×10^8	8.4×10^{-11}	0.93

mutations on chromosome II; it may be an underestimate because very slightly deleterious mutations were missed, but it is unlikely to be an overestimate.

The micro-organism rates are based on the rate at which specific genes mutate to a non-functional allele, scaled up by the known amount of DNA per genome. The scaling procedure is permissible in prokaryotes, in which a large part of the DNA consists of functional genes. But the question arises whether the method may not seriously underestimate the total rate, because mutations which alter but do not destroy gene function are missed, although they might be selectively important in nature. It is possible to investigate this by considering 'chain-terminating' mutations. Approximately 5% of all base substitution mutations are chain terminators, and all these are likely to destroy gene function. It is possible, for some genes, to estimate what fraction of the discovered mutations are chain terminators, because protein fragments are produced. In this way it can be shown that the estimates in the table, although perhaps underestimates, are certainly not out by an order of magnitude. It follows that the various values in the table are, at least roughly, estimates of the same thing, namely the total rate of deleterious mutations per genome.

Drake (1974) argues that the uniformity of the rates per genome in micro-organisms implies some kind of optimising selection. This is reasonable, although it is hard to suggest a mechanism. One difficulty is that we do not know how important recombination is in the evolution of prokaryotes. It is equally difficult to explain the much higher rate per genome in *Drosophila*. In one sense, this rate merely reflects the larger number of gene replications per generation and the larger amount of meaningful DNA per genome. It may be that a value of 10^{-10} per base pair replication is close to the lowest limit which can be achieved without excessive cost, but this is mere speculation.

Cox (1976) points out that in *E. coli* the spontaneous rate per base pair replication is of the order 10^{-10}, and that a number of mutants are known with values of the order 10^{-5} to 10^{-6}. He suggests that the latter value corresponds to that which would be expected simply from the energy levels associated with complementary base pairing. The level of 10^{-10} depends on 'proof-reading' after replication (i.e. the detection immediately after replication of any errors which have been made, and their correction). Thus the mutants represent a total loss of the proof-reading function.

Cox is puzzled by the fact that the spontaneous rate has been reduced to the very low level observed, because of the low selection pressure involved. He suggested that, since a gene responsible for DNA replication replicates itself, this could lead to the evolution of a very low error rate by mutational balance, without selective differences. After some correspondence, Dr Cox and I now agree that there is a fallacy in his original argument on this point. It may be that he underestimated the effectiveness, in the long run, of even very weak selective forces.

However, his paper does raise the problem of how the rate could evolve from the 'thermodynamic' level of approximately 10^{-6} to the 'proof-read' level of 10^{-10}. Clearly this could not have happened in a single step. As in the evolution of other adaptations, there must have been a series of intermediate steps, each bringing an improvement in accuracy. Hence it should be possible to find mutator genes which cause smaller reductions in accuracy than those so far found.

With an error rate of 10^{-6} per base pair replication, an organism could have an upper limit of approximately 1000 genes. This upper limit is possible only if the organism has genetic recombination, or if the majority of errors are immediately lethal or almost so. If most errors produce only a slight reduction in fitness (to $1 - s$, where s is small), and if there is no recombination, then Muller's ratchet (p. 35) will operate when the number of genes is greater than approximately $1000s$.

This raises in a particularly clear way the 'boot-strapping' problem of the origin of life. If an organism has both recombination and proof-reading, it can evolve a large genome. But these functions require enzymes, which must be programmed by genes. So an organism cannot have a large genome without recombination and proof-reading, and it cannot have recombination and proof-reading without a large genome. However, the problem may not be too severe. A thousand genes should be quite adequate to programme the proof-reading and recombination functions and leave plenty over to cope with metabolic activities. Serious difficulties would arise only if a large proportion of mutations cause slightly deleterious effects; if so, Muller's ratchet would operate when the genome was still too small to include genes reducing the mutation rate.

B   Mutation and hitch-hiking

It may be that the decisive step towards an understanding of the evolution of mutation rates was an experiment rather than a theory. The experiment of Cox & Gibson (1974), which showed that there can be selection for a mutator gene in an asexual population of bacteria, was discussed in Chapter 7, section A. The essential point is that a gene causing a favourable mutation is carried to fixation by the mutation it has caused. The question, of course, is whether this hitch-hiking process can be important in the presence of recombination.

Following up his earlier argument that selection for increased mutation can occur only in asexual populations, Leigh (1973) enquired whether, in a sexual population, a mutator gene might increase in frequency because of favourable mutations in the region of chromosome closely linked to it. He concluded that the effect would be vanishingly small. However, he may have underestimated the effect, because he analysed the process in an infinite population. In a population whose size is less than the reciprocal of the per base mutation rate (approximately, $< 10^{10}$), new favourable mutations will arise in coupling with the mutator gene which caused them (or, as we should perhaps say, which permitted them to happen). Leigh is certainly right in thinking that selection for higher mutation rates will be much stronger in the absence of recombination, but the hitch-hiking effect may still be significant even in a sexual population.

In conclusion, it is instructive to compare the arguments concerning the evolution of recombination and mutation rates. In both, there is the possibility of a balance between short-term selection for reduced rate and long-term inter-group selection for an increased one. In both, there is the possibility that selection on the genetic consequences is always for a reduced rate, and that this is balanced by selection arising from the physiological costs – in reduced replication rate or in a disturbed meiosis – of a reduction; this seems a very real possibility in the case of the mutation rate in higher organisms. In both, the crucial process may be the hitch-hiking effect of favourable mutations or recombinants on the frequency of the gene which caused them. Finally, we have always to remember that mutation and recombination are both, in their origin, aspects of the same basic processes of replication and repair of the genetic material.

Appendix: The mutation rate in an infinite asexual population

Consider an infinite, asexual haploid population. The environment changes so that at intervals of n generations an allele which was previously unfavourable becomes selectively advantageous; let two such successive mutations be $a \to A$ and $b \to B$. The population contains a pair of alleles M, m affecting the mutation rate. The mutation rates per generation to A or B are u in M and u' in m (it would not alter the conclusions if the rates to A and B were different, provided they remained in the ratio $u : u'$). The rate of deleterious mutations per generation is $(1 - u)^k = e^{-uk}$ in M and $e^{-u'k}$ in m; k is a large number (roughly, it is the number of mutational sites per genome).

In generation $t = 0$, allele A becomes favourable. Because the population is infinite, both AM and Am genotypes already exist in mutational equilibrium. If p is the frequency of M in the whole population, then

$$\frac{\text{frequency of } AM}{\text{frequency of } Am} = \frac{pu}{(1 - p)u'}.$$

n generations later, allele B becomes favourable. At this time,

$$\frac{\text{frequency of } AM}{\text{frequency of } Am} = \frac{pue^{-ukn}}{(1 - p)u'e^{-u'kn}} = \frac{P}{1 - P} \quad \text{say,}$$

where P is the frequency of AM in the A sub-population. After this time, allele B increases in frequency, but since the population is asexual we are only interested in those B mutants which occur in the A sub-population. Hence

$$\frac{\text{frequency of } ABM}{\text{frequency of } ABm} = \frac{Pu}{(1-P)u'},$$

and we are starting a new cycle of n generations, with the frequency of M changed from p to P.

An allele M which maximises ue^{-ukn} will increase in frequency relative to any other allele. Hence

$$\mathrm{d}(ue^{-ukn})/\mathrm{d}u = e^{-ukn} - kune^{-ukn} = 0,$$

or $ku = 1/n$.

Now $ku = M$, the mutation rate per genome, and $1/n = E$, the allelic substitution rate. So, for this model,

$$M = E.$$

This result was obtained by Leigh (1970) for an asexual population. It agrees with the optimal mutation rate obtained by Kimura (1967) based on minimising the total genetic load, which is what would be expected from inter-group selection. The correspondence is to be expected, since selection in an asexual population is in effect inter-clonal selection.

References

Abdullah, N.F. & Charlesworth, B. (1974). Selection for reduced crossing over in *Drosophila melanogaster*. *Genetics* **76:** 447–451.

Acton, A.B. (1961). An unsuccessful attempt to reduce recombination by selection. *Am. Natur.* **95:** 119–120.

Adams, M.S. & Neel, J.V. (1967). Children of incest. *Pediatrics* **40:** 55–62.

Alexander, R.D. (1977*a*). Natural selection and the analysis of human sociality. In *Changing Scenes in Natural Sciences,* ed. C.E. Goulden. Philadelphia Academy of Natural Sciences.

Alexander, R.D. (1977*b*). Evolution, human behaviour and determinism. *Proceedings of the 1976 Biennial Meeting of the Philosophy of Science Association,* Vol. 2.

Alexander, R.D. & Sherman, P.W. (1977). Local mate competition and parental investment in social insects. *Science* **196:** 494–500.

Allard, R.W. (1963). Evidence for genetic restriction of recombination in the lima bean. *Genetics* **48:** 1389–1395.

Allard, R.W., Kahler, A.L. & Clegg, M.T. (1977). Estimation of mating cycle components of selection in plants. In *Measuring Selection in Natural Populations,* ed. F.B. Christiansen & T.M. Fenchel, pp. 1–19. Berlin: Springer-Verlag.

Amadon, D. (1959). The significance of sexual difference in size among birds. *Proc. Am. Phil. Soc.* **103:** 531–536.

Assem, J. Van Den (1971). Some experiments on sex ratio and sex regulation in the Pteromalid *Lariophagus distinguendus. Neth. Jl Zool.* **21:** 373–402.

Babcock, E.B. & Stebbins, G.L. (1938). The American species of *Crepis:* their relationships and distribution as affected by polyploidy and apomixis. *Carnegie Institute, Washington, Publ.* No. 504.

Bacci, G. (1965). *Sex Determination.* London: Pergamon Press.

Baker, H.G. (1955). Self-compatibility and establishment after 'long-distance' dispersal. *Evolution* **9:** 347–348.

Baker, H.G. (1959). Reproductive methods as factors in speciation in flowering plants. *Cold Spring Harbor Symp. quant. Biol.* **24:** 177–191.

Baker, H.G. (1967). Support for Baker's law – as a rule. *Evolution* **21:** 853–856.

Banta, A.M. (1925). A thelytokous race of *Cladocera* in which pseudosexual reproduction occurs. *Z. indukt. Abstamm. Vererbungsl.* **40:** 28–41.

Bar-Anon, R. & Robertson, A. (1975). Variation in sex ratio between progeny groups in dairy cattle. *Theoret. app. Genet.* **46:** 63–65.

Basrur, V.R. & Rothfels, K.H. (1959). Triploidy in natural populations of the black fly *Crephia mutata* (Mallock). *Can. Jl Zool.* **37:** 571–589.

Bateman, A.J. (1948). Intrasexual selection in *Drosophila. Heredity* **2:** 349–368.

Bawa, K.S. & Opler, P.A. (1975). Dioecism in tropical forest trees. *Evolution* **29:** 167–179.

Bell, G. (1978). The evolution of anisogamy. *J. theoret. Biol.* (in press).

Bellrose, F.C., Scott, T.G., Hawkins, A.S. & Low, J.B. (1961). Sex ratios and age ratios in North American Ducks. *Bull. Ill. nat. Hist. Surv.* **27:** 391–474.

Bengtsson, B.O. (1978). Avoid inbreeding: at what cost? *J. theoret. Biol.* (in press).

Bergerard, J. (1962). Parthenogenesis in the Phasmidae. *Endeavour* **21:** 137–143.

Birky, C.W. & Gilbert, J.J. (1971). Parthenogenesis in rotifers: the control of sexual and asexual reproduction. *Am. Zool.* **11:** 245–266.

Bodmer, W.F. (1970). The evolutionary significance of recombination in prokaryotes. *Symp. Soc. gen. Microbiol.* **20:** 279–294.

Bodmer, W.F. & Felsenstein, J. (1967). Linkage and selection: theoretical analysis of the deterministic two locus random mating model. *Genetics* **57:** 237–265.

Breder, C.M. & Rosen, D.E. (1966). *Modes of Reproduction in Fishes.* New York: Natural History Press.

Brues, A.M. (1964). The cost of evolution vs. the cost of not evolving. *Evolution* **18:** 379–383.

Bruning, D.F. (1973). The greater rhea chick and egg delivery route. *Nat. Hist.* **82:** 68–75.

Bulmer, M.G. (1973) Inbreeding in the great tit. *Heredity* **30:** 313–325.

Carlquist. S. (1965). *Island Life.* New York: Natural History Press.

Carlquist, S. (1966). The biota of long distance dispersal. IV. Genetic systems in the flora of oceanic islands. *Evolution* **20:** 433–455.

Carson, H.L. (1967). Selection for parthenogenesis in *Drosophila mercatorium. Genetics* **55:** 157–171.

Catcheside, D.G. (1968). The control of recombination in *Neurospora crassa.* In *Replication and Recombination of Genetic Material,* ed. W.J. Peacock & R.D. Brock, pp. 216–226. Canberra: Australian Academy of Science.

Catcheside, D.G. (1975). Occurrence in wild strains of *N. crassa* of genes controlling recombination. *Aust. J. biol. Sci.* **28:** 213–225.

Charlesworth, B. (1976). Recombination modification in a fluctuating environment. *Genetics* **83:** 181–195.

Charlesworth, B. (1977). Population genetics, demography and the sex ratio.

In *Measuring Selection in Natural Populations*, ed. F.B. Christiansen & T.M. Fenchel, pp. 345–363. Berlin: Springer-Verlag.

Charlesworth, B. & Charlesworth, D. (1973). Selection of new inversions in multi-locus genetic systems. *Genet. Res., Camb.* **23:** 167–183.

Charlesworth, B. & Charlesworth, D. (1976*a*). An experiment on recombination load in *Drosophila melanogaster. Genet. Res., Camb.* **25:** 267–274.

Charlesworth, D. & Charlesworth, B. (1976*b*). Theoretical genetics of Batesian mimicry. II. Evolution of supergenes. *J. theoret. Biol.* **55:** 305–324.

Charlesworth, D., Charlesworth, B. & Strobeck, C. (1977). Effects of selfing on selection for recombination. *Genetics* **86:** 213–226.

Charnov, E.L. (1978*a*). The genetical evolution of patterns of sexuality. I. Darwinian fitness. *Am. Natur.* (in press).

Charnov, E.L. (1978*b*). Sex ratio selection in eusocial Hymenoptera. *Am. Natur.* (in press).

Charnov, E.L. & Bull, J. (1977). When is sex environmentally determined? *Nature, Lond.* **266:** 828–830.

Charnov, E.L., Maynard Smith, J. & Bull, J.J. (1976). Why be an hermaphrodite? *Nature, Lond.* **263:** 125–126.

Chewyreuv, I. (1913). Le rôle des femelles dans la détermination du sexe de leur descendance dans le groupe des Ichneumonides. *C.r. Séanc. Soc. Biol.* **74:** 695.

Chinnici, J.P. (1971*a*). Modification of recombination frequency in *Drosophila*. I. Selection for increased and decreased crossing over. *Genetics* **69:** 71–83.

Chinnici, J.P. (1971*b*). Modification of recombination frequency in *Drosophila*. II. The polygenic control of crossing over. *Genetics* **69:** 85–96.

Christensen, B. (1961). Studies on cytotaxonomy and reproduction in the Enchytraeidae. With notes on parthenogenesis and polyploidy in the animal kingdom. *Hereditas* **47:** 387–450.

Clarke, B. (1972). Density-dependent selection. *Am. Natur.* **106:** 1–13.

Clarke, C.A. & Sheppard, P.M. (1971). Further studies on the genetics of the mimetic butterfly *Papilio memnon* L. *Phil. Trans. R. Soc. B* **263:** 35–70.

Clayton, G.A. & Robertson, A. (1957). An experimental check on quantitative genetical theory. II. The long-term effects of selection. *J. Genet.* **55:** 152–170.

Clutton-Brock, T.H. & Harvey, P.H. (1977). Primate ecology and social organisation. *J. Zool.* **183:** 1–39.

Cox, E.C. (1976). Bacterial mutator genes and the control of spontaneous mutation. *A. Rev. Genet.* **10:** 135–156.

Cox, E.C. & Gibson, T.C. (1974). Selection for high mutation rates in chemostats. *Genetics* **77:** 169–184.

198   *References*

Crosby, J.L. (1949). Selection of an unfavourable gene complex. *Evolution* **3:** 212–230.

Crow, J.F. & Kimura, M. (1965). Evolution in sexual and asexual populations. *Am. Natur.* **99:** 439–450.

Crow, J.F. & Kimura, M. (1970). *An Introduction to Population Genetics Theory.* New York: Harper and Row.

Cuellar, O. (1971). Reproduction and the mechanism of meiotic restitution in the parthenogenetic lizard, *Cnemidophorus uniparens. J. Morphol.* **133:** 139–165.

Cuellar, O. (1974). The origin of parthenogenesis in vertebrates: the cytogenetic factors. *Am. Natur.* **108:** 625–648.

Cuellar, O. (1976). Intraclonal histocompatibility in a parthenogenetic lizard: evidence of genetic homogeneity. *Science,* **193:** 150–153.

Darlington, C.D. (1939). *The Evolution of Genetic Systems.* Cambridge University Press.

Darwin, C. (1871). *The Descent of Man and Selection in Relation to Sex.* London: John Murray.

Darwin, C. (1876). *The Effects of Cross and Self Fertilisation in the Vegetable Kingdom,* second edition, pp. 414–415. London: John Murray.

Darwin, C. (1877). *The Different Forms of Flowers on Plants of the Same Species.* London: John Murray.

Davis, G.W.F. & O'Donald, P. (1976). Sexual selection for a handicap: a critical analysis of Zahavi's model. *J. theoret. Biol.* **57:** 345–354.

Dawkins, R. (1976). *The Selfish Gene.* Oxford University Press.

Dawkins, R. & Carlisle, T.R. (1976). Parental investment, mate desertion and a fallacy. *Nature, Lond.* **262:** 131–133.

Detlefsen, J.A. & Roberts, E. (1921). Studies on crossing over. I. The effect of selection on crossover values. *J. exptl Zool.* **32:** 333–354.

Dewees, A.A. (1970). Two-way selection for recombination rates in *Tribolium castaneum.* (Abstract). *Genetics* **64:** 516–517.

Dobzhansky, Th. (1937). *Genetics and the Origin of Species.* Columbia University Press.

Drake, J.W. (1974). The role of mutation in bacterial evolution. *Symp. Soc. gen. Microbiol.* **24:** 41–58.

✓ Drickamer, L.C. & Vessey, S.H. (1973). Group changing in free-ranging male rhesus monkeys. *Primates* **14:** 359–368.

Edwards, A.W.F. (1962). Genetics and the human sex ratio. *Adv. Genet.* **11:** 239–272.

Edwards, A.W.F. (1966). Sex ratio data analysed independently of family limitation. *Ann. Hum. Genet., Lond.* **29:** 337–347.

Edwards, A.W.F. (1970). The search for genetic variability of the sex ratio. *J. biosoc. Sci., Suppl.* **2,** 55–60.

Ehrman, L. (1972). Genetics and sexual selection. In *Sexual Selection and the Descent of Man*, ed. B. Campbell, pp. 105–135. Chicago: Aldine.

Emerson, A.E. (1960). The evolution of adaptation in population systems. In *Evolution after Darwin*, vol. 1, *The Evolution of Life*, ed. S. Tax, pp. 307–348. University of Chicago Press.

Eshel, I. & Feldman, M.W. (1970). On the evolutionary effect of recombination. *Theoret. Pop. Biol.* **1**: 88–100.

Ernst, A. (1936). Heterostylie-forschung Versuche zur genetischen Analyse eines Organisations – und 'Anpassungs' merkmales. *Z. indukt. Abstamm. Vererbungsl.* **71**: 156–230.

Falconer, D.S. (1954). Selection of sex ratio in mice and *Drosophila*. *Am. Natur.* **88**, 385–397.

Falconer, D.S. (1960). *Introduction to Quantitative Genetics*. Edinburgh: Oliver and Boyd.

Feldman, M.W., Franklin, I. & Thomson, G. (1974). Selection in complex genetic systems. I. The symmetric equilibria of the three locus symmetric viability model. *Genetics* **76**: 135–162.

Felsenstein, J. (1971). On the biological significance of the cost of gene substitution. *Am. Natur.* **105**: 1–11.

Felsenstein, J. (1974). The evolutionary advantage of recombination. *Genetics* **78**: 737–756.

Felsenstein, J. & Yokoyama, S. (1976). The evolutionary advantage of recombination. II. Individual selection for recombination. *Genetics* **83**: 845–859.

Fisher, R.A. (1930). *The Genetical Theory of Natural Selection*. Oxford University Press.

✓ Frame, L.H. & Frame, G.W. (1976). Female African wild dogs emigrate. *Nature, Lond.* **263**: 227–229.

Franklin, I. & Lewontin, R.C. (1970). Is the gene the unit of selection? *Genetics* **65**: 707–734.

Freeman, D.C., Klikoff, L.G. & Harper, K.T. (1976). Differential resource utilization by the sexes of dioecious plants. *Science* **193**: 597–599.

Frith, H.J. (1962). *The Mallee-Fowl*. Sydney: Angus and Robertson.

Ghiselin, M.T. (1969). The evolution of hermaphroditism among animals. *Q. Rev. Biol.* **44**: 189–208.

Ghiselin, M.T. (1974). *The Economy of Nature and the Evolution of Sex.* University of California Press.

Gilbert, J.L. (1974). Dormancy in rotifers. *Trans. Am. Micros. Soc.* **93**: 490–513.

Gilpin, M.E. (1975). *Group Selection in Predator–Prey Communities*. Princeton University Press.

Glesener, R.R. & Tilman, D. 1978. Sexuality and the components of environmental uncertainty. *Am. Natur.* (in press).

Godley, E.J. (1955). Monoecy and incompatibility. *Nature, Lond.* **176:** 1176–1177.

Grant, V. (1958). The regulation of recombination in plants. *Cold Spring Harbour Symp. quant. Biol.* **23:** 337–363.

Greene, P.J. (1978). Promiscuity, paternity and culture. *Am. Ethnol.* (in press).

Greenwood, P.J. & Harvey, P.H. (1976). The adaptive significance of variation in breeding area fidelity of the blackbird (*Turdus merula* L.). *J. Anim. Ecol.* **45:** 887–898.

Greenwood, P.J. & Harvey, P.H. (1977). Feeding strategies and dispersal of territorial passerines: a comparative study of the blackbird, *Turdus merula,* and the greenfinch, *Carduelis chloris. Ibis* **119:** 528–531.

Haas, R. (1976). Sexual selection in *Notobranchius guentheri* (Pisces: Cyprinodontidae). *Evolution* **30:** 614–622.

Haldane, J.B.S. (1957). The cost of natural selection. *J. Genet.* **55:** 511–524.

Hamilton, W.D. (1964). The genetical theory of social behaviour. I and II. *J. theoret. Biol.* **7:** 1–16; 17–32.

Hamilton, W.D. (1967). Extraordinary sex ratios. *Science* **156:** 477–488.

✓ Harcourt, A.H., Stewart, K.S. & Fossey, D. (1976). Male emigration and female transfer in the wild mountain gorilla. *Nature, Lond.* **263:** 226–227.

Harper, J.L., Lovell, P.H. & Moore, K.G. (1970). The shapes and sizes of seeds. *A. Rev. ecol. Syst.* **1:** 327–356.

Haskins, C.P., Haskins, E.F., McLaughlin, J.J.A. & Hewitt, R.E. (1961). Polymorphism and population structure in *Lebistes reticulatus,* an ecological study. In *Vertebrate Speciation,* ed. W.F. Blair. University of Texas Press.

Hebert, P.D.N. (1974). Enzyme variability in natural populations of *Daphnia magna.* II. Genotypic frequencies in permanent populations. *Genetics,* **77:** 323–334. III. Genotypic frequencies in intermittent populations. *Ibid:* 335–344.

Hickey, W.A. & Craig, G.P. (1966). Genetic distortion of sex ratio in a mosquito, *Aedes aegypti. Genetics* **53:** 1177–1196.

Hill, J.L. (1974). *Peromyscus:* effect of early pairing on reproduction. *Science* **186:** 1042–1044.

Hill, W.G. & Robertson, A. (1966). The effect of linkage on limits to artificial selection. *Genet. Res., Camb.* **8:** 269–294.

Hollingsworth, M.J. & Maynard Smith, J. (1955). The effects of inbreeding on rate of development and on fertility in *Drosophila subobscura. J. Genet.* **53:** 295–314.

Howe, H.F. (1977). Sex ratio adjustment in the common grackle. *Science* **198:** 744–746.

Hutchinson, G.E. (1967). *A Treatise on Limnology.* New York: Wiley.

Jenni D.A. (1974). Evolution of polyandry in birds. *Am. Zool.* **14**: 129–144.

Jenni, D.A. & Collier, G. (1972). Polyandry in the American jacana (*Jacana spinosa*). *Auk* **89**: 743–765.

Jinks, J.L. & Mather, K. (1955). Stability in development of heterozygotes and homozygotes. *Proc. R. Soc.* B **143**: 561–578.

Karlin, S. (1973). In *The Mathematical Theory of the Dynamics of Biological Populations,* eds. M. S. Bartlett & R. W. Hiorns. Academic Press, London.

Karlin, S. & Feldman, M.W. (1970). Linkage and selection: two locus symmetric viability models. *Theoret. Pop. Biol.* **1**: 39–71.

Karlin, S. & McGregor, J. (1972). Polymorphisms for genetic and ecological systems with weak coupling. *Theoret. Pop. Biol.* **3**: 210–238.

Kear, J. (1970). The adaptive radiation of parental care in waterfowl. In *Social Behaviour in Birds and Mammals,* ed. J.H. Crook, pp. 357–392. London: Academic Press.

Kerfoot, W.C. (1974). Egg-size cycle of a cladoceran. *Ecology* **55**: 1259–1270.

Kidwell, M.G. (1972*a*). Genetic changes of recombination value in *Drosophila melanogaster*. I. Artificial selection for high and low recombination and some properties of recombination-modifying genes. *Genetics* **70**: 419–432.

Kidwell, M.G. (1972*b*). Genetic changes of recombination value in *Drosophila melanogaster*. II. Simulated natural selection. *Genetics* **70**: 433–443.

Kimura, M. (1956). A model of a genetic system which tends to closer linkage by natural selection. *Evolution* **10**: 278–287.

Kimura, M. (1960). Optimum mutation rate and degree of dominance as determined by the principle of minimum genetic load. *J. Genet.* **57**: 21–34.

Kimura, M. (1967). On the evolutionary adjustment of spontaneous mutation rates. *Genet. Res.* **9**: 23–34.

Kimura, M. (1968). Evolutionary rate at the molecular level. *Nature, Lond.* **217**: 624–626.

Kimura, M. & Ohta, T. (1971). *Theoretical Aspects of Population Genetics.* Princeton University Press.

King, J.L. (1967). Continuously distributed factors affecting fitness. *Genetics* **55**: 483–492.

Knowlton, N. (1974). A note on the evolution of gamete dimorphism. *J. theoret. Biol.* **46**: 283–285.

Knox, P.B. (1967). Apomixis: seasonal and population differences in a grass. *Science* **157**: 325–326.

Lack, D. (1968). *Ecological Adaptations for Breeding in Birds.* London: Methuen.

Langley, C.H. (1977). Nonrandom associations between allozymes in natural populations of *Drosophila melanogaster*. In *Measuring Selection in Natural Populations*, ed. F.B. Christiansen & T.M. Fenchel, pp. 265–273. Berlin: Springer-Verlag.

Leigh, E.G. (1970). Natural selection and mutability. *Am. Natur.* **104:** 301–305.

Leigh, E.G. (1973). The evolution of mutation rates. *Genetics, Suppl.* **73:** 1–18.

Lerner, I.M. (1954). *Genetic Homeostasis*. Edinburgh: Oliver and Boyd.

Levene, H. (1953). Genetic equilibrium when more than one ecological niche is available. *Am. Natur.* **87:** 131–133.

Levin, D.A. (1975). Pest pressure and recombination systems in plants. *Am. Natur.* **109:** 437–451.

Levins, R. (1968). *Evolution in Changing Environments*. Princeton University Press.

Levi-Strauss, C. (1968). *Structural Anthropology*, English edition. London: Allen Lane.

Lewis, D. (1941). Male sterility in natural populations of hermaphrodite plants. *New Phytol.* **40:** 56–63.

Lewis, K.R. & John, B. (1963). *Chromosome Marker*. London: Churchill.

Lewontin, R.C. (1971). The effect of genetic linkage on the mean fitness of a population. *Proc. natn. Acad. Sci. USA* **68:** 984–986.

Lewontin, R.C. (1974). *The Genetic Basis of Evolutionary Change*. Columbia University Press.

Lewontin, R.C. & Hubby, J.L. (1966). A molecular approach to the study of genic heterozygosity in natural populations. II. Amount of variation and degree of heterozygosity in natural populations of *Drosophila pseudoobscura*. *Genetics* **54:** 595–609.

Lewontin, R.C. & Kojima, K. (1960). The evolutionary dynamics of complex polymorphisms. *Evolution* **14:** 458–472.

Lokki, J., Suomalainen, E., Saura, A. & Lankinen, P. (1975). Genetic polymorphism and evolution in parthenogenetic animals. II. Diploid and polyploid *Solenobia triquetrella* (Lepidoptera: Psychidae). *Genetics* **79:** 513–525.

McAlister, W.H. (1972). Variation in *Rana pipiens* in Texas. *Am. Midl. Natur.* **67:** 334–363.

MacArthur, R.H. (1965). Ecological consequences of natural selection. In *Theoretical and Mathematical Biology*, ed. T. Waterman & H. Morowitz, pp. 388–397. Blaisdell: New York.

MacArthur, R.H. & Wilson E.O. (1967). *Island Biogeography*. Princeton University Press.

Marshall, D.R. & Allard, R.W. (1970). Maintenance of isozyme polymorphism in natural populations of *Avena sativa*. *Genetics* **66:** 393–399.

Maslin, P.T. (1971). Parthenogenesis in reptiles. *Am Zool.* **11:** 361–380.

Mather, K. (1943). Polygenic inheritance and natural selection. *Biol. Rev.* **18:** 32–64.

Mather, K. & Harrison, B.S. (1949). The manifold effect of selection. *Heredity* **3:** 1–52.

Maynard Smith, J. (1956). Fertility, mating behaviour and sexual selection in *Drosophila subobscura*. *J. Genet.* **54:** 261–279.

Maynard Smith, J. (1962). Disruptive selection, polymorphism and sympatric speciation. *Nature, Lond.* **195:** 60–62.

Maynard Smith, J. (1968a). Evolution in sexual and asexual populations. *Am. Natur.* **102:** 469–473.

Maynard Smith, J. (1968b). 'Haldane's dilemma' and the rate of evolution. *Nature, Lond.* **219:** 1114–1116.

Maynard Smith, J. (1971a). What use is sex? *J. theoret. Biol.* **30:** 319–335.

Maynard Smith, J. (1971b). The origin and maintenance of sex. In *Group Selection*, ed. G.C. Williams, pp. 163–175. Chicago: Aldine–Atherton.

Maynard Smith, J. (1974). Recombination and the rate of evolution. *Genetics* **78:** 299–304.

Maynard Smith, J. (1976a). A comment on the Red Queen. *Am. Natur.* **110:** 325–330.

Maynard Smith, J. (1976b). What determines the rate of evolution? *Am. Natur.* **110:** 331–338.

Maynard Smith, J. (1976c). Group selection. *Q. Rev. Biol.* **51:** 277–283.

Maynard Smith, J. (1976d). Sexual selection and the handicap principle. *J. theoret. Biol.* **57:** 239–242.

Maynard Smith, J. (1976e). Evolution and the theory of games. *Am. Scient.* **64:** 41–45.

Maynard Smith, J. (1976f). A short term advantage for sex and recombination through sib-competition. *J. theoret. Biol.* **63:** 245–258.

Maynard Smith, J. (1977a). Parental investment – a prospective analysis. *Anim. Behav.* **25:** 1–9.

Maynard Smith, J. (1977b). The sex habit in plants and animals. In *Measuring Selection in Natural Populations*, ed. F.B. Christiansen & T.M. Fenchel, pp. 265–273. Berlin: Springer-Verlag.

Maynard Smith, J. & Haigh, J. (1974). The hitch-hiking effect of a favourable gene. *Genet. Res., Camb.* **23:** 23–35.

Maynard Smith, J. & Price, G.R. (1973). The logic of animal conflict. *Nature, Lond.* **246:** 15–18.

Milkman, R.D. (1967). Heterosis as a major cause of heterozygosity in nature. *Genetics* **55:** 493–495.

Mockford, E.L. (1971). Parthenogenesis in psocids (Insecta: Psoceptera). *Am. Zool.* **11:** 327–339.

Moore, W.S. (1976). Components of fitness in the unisexual fish *Poeciliopsis monacha-occidentalis*. *Evolution* **30:** 564–578.

Moran, P.A.P. (1962). *The Statistical Processes of Evolutionary Theory.* Oxford University Press.

Mukai, T., Chigusa, S.T., Mettler, L.E. & Crow, J.F. (1972). Mutation rate and dominance of genes affecting viability in *Drosophila melanogaster. Genetics* **72**: 335–355.

Mukai, T. & Yamaguchi, O. (1974). The genetic structure of natural populations of *Drosophila melanogaster.* XI. Genetic variability in a local population. *Genetics* **76**: 339–366.

Muller, H.J. (1932). Some genetic aspects of sex. *Am. Natur.* **66**: 118–138.

Muller, H.J. (1964). The relation of recombination to mutational advance. *Mutat. Res.* **1**: 2–9.

Muntzing, A. (1958). The balance between sexual and apomictic reproduction in some hybrids of *Potentilla. Hereditas* **44**: 145–160.

Murray, J. (1964). Multiple mating and effective population size in *Cepaea nemoralis. Evolution* **18**: 283–291.

Murray, J. (1975). The genetics of the Mollusca. In *Handbook of Genetics,* vol. 3, ed. R.C. King. New York, Plenum.

Neaves, W.B. (1969). Adenosine deaminase phenotypes among sexual and parthenogenetic lizards in the genus *Cnemidophorus* (Teiidae). *J. exptl Zool.* **171**: 175–184.

Nei, M. (1967). Modification of linkage intensity by natural selection. *Genetics* **57**: 625–641.

Newton, I. & Marquiss, M. (1978). Sex ratio among nestlings of the European sparrowhawk. *Am. Natur.* (in press).

Nikolskii, G.V. (1969). *Fish Population Dynamics.* Edinburgh: Oliver and Boyd.

Novitski, E. & Kimball, A.W. (1958). Birth order, parental ages and sex of offspring. *Am. J. hum. Genet.* **21**: 123–131.

Nur, U. (1971). Parthenogenesis in coccids (Homoptera). *Am. Zool.* **11**: 301–308.

O'Donald, P. (1962). The theory of sexual selection. *Heredity* **17**: 541–552.

O'Donald, P. (1976). Mating preferences and their genetic effects in models of sexual selection for colour phases of the Arctic Skua. In *Population Genetics and Ecology,* ed. S. Karlin & E. Nevo, pp. 411–430. New York: Academic Press.

Olsen, M.W. (1965). Twelve year summary of selection for parthenogenesis in the Beltsville small white turkey. *Br. Poultry Sci.* **6**: 1–6.

Orians, G.H. (1969). On the evolution of mating systems in birds and mammals. *Am. Natur.* **103**: 589–603.

Packer, C. (1975). Male transfer in olive baboons. *Nature, Lond.* **255**: 219–220.

Parker, E.D. & Selander, R.K. (1976). The organization of genetic diversity in the parthenogenetic lizard *Cnemidophorus tesselatus*. *Genetics* **84:** 791–805.

Parker, G.A. (1974). Courtship persistence and female guarding as male time investment strategies. *Behavior* **48:** 157–184.

Parker, G.A. (1978). Selection on non-random fusion of gametes during the evolution of anisogamy. *J. theoret. Biol.* (in press).

Parker, G.A., Baker, R.R. & Smith, V.G.F. (1972). The origin and evolution of gamete dimorphism and the male–female phenomenon. *J. theoret. Biol.* **36:** 529–553.

Pijnacker, L.P. (1967). Oogenesis in the parthenogenetic stick insect *Sipyloidae sipylus* Westwood (Orthoptera, Phasmidae). *Genetica* **38:** 504–515.

Pijnacker, L.P. (1969). Automictic parthenogenesis in the stick insect *Bacillus rossius* Rossi (Chelentoptera, Phasmidae). *Genetica* **40:** 393–399.

Pimentel, R.A. (1959). Positive embryo–maternal size correlation in the northern alligator lizard, *Gerrhonotus caeruleus principis*. *Herpetologica* **15:** 6–8.

Poulsen, D.F. & Sakaguki, B. (1961). Nature of sex ratio agent in *Drosophila*. *Science* **133:** 1489–1490.

Preston, F.W. (1962). The canonical distribution of commonness and rarity. I and II. *Ecology* **43:** 185–214, 410–432.

Pusey, A.E. (1977). Intercommunity transfer of chimpanzees in Gombe National Park. In *Perspectives on Human Evolution*, vol. 5, *Great Apes*, ed. D.A. Hamburg. London and New York: Benjamin (in press).

Ralls, K. (1976). Mammals in which females are larger than males. *Q. Rev. Biol.* **51:** 245–276.

Rees, H. (1956). Heterosis in chromosome behaviour. *Proc. R. Soc.* B **144:** 150–159.

Rensch, B. (1960). *Evolution Above the Species Level*. Columbia University Press.

Richards, A.J. (1973). The origin of *Taraxacum* agamo-species. *Bot. J. Linn. Soc.* **66:** 189–211.

Robertson, J.G. (1966). The chromosomes of bisexual and parthenogenetic species of *Calligrapha* (Coleoptera: Chrysomelidae) with notes on sex ratio, abundance and egg number. *Can. J. genet. Cytol.* **8:** 695–732.

Ruttner-Kolisko, A. (1946). Über das Auftreten unbefruchteter Dauereier bei Anuraea aculeata (*Keratella quadreta*). *Öst. zool. Z.* **1:** 179–191.

Schull, W.J. & Neel, J.V. (1965). *The Effects of Inbreeding on Japanese Children*. New York: Harper and Row.

Schultz, R.J. (1969). Hybridization, unisexuality and polyploidy in the teleost, *Poeciliopsis* (Poeciliidae) and other vertebrates. *Am. Natur.* **103:** 605–619.

Schultz, R.J. (1973). Unisexual fish: laboratory synthesis of a 'species'. *Science* **179:** 180–181.

Scudo, F.M. (1964). Sex population genetics. *Ric. Sci.* **34:** 93–146.

Seiler, J. (1963). Untersuchungen über die Enstehung der Parthenogenese bei *Solenobia triquetrella* F. R. (Lepidoptera Psychidae). *Z. VererbLehre* **94:** 29–66.

Selander, R.K. (1972). Sexual selection and dimorphism in birds. *Sexual Selection and the Descent of Man,* ed. B. Campbell, pp. 180–230. Chicago: Aldine.

Selander, R.K. & Kaufman, D.W. (1973). Self-fertilization and genetic population structure in a colonizing land snail. *Proc. natn. Acad. Sci. USA* **70:** 1186–1190.

Shaw, D.D. (1971). Genetic and environmental components of chiasma control. I. Spatial and temporal variation in *Schistocerca* and *Stethophyma*. *Chromosoma* **34:** 281–301.

Shaw, D.D. (1972). Genetic and environmental components of chiasma control. II. The response to selection in *Schistocerca*. *Chromosoma* **37:** 297–308.

Shaw, D.D. (1974). Genetic and environmental components of chiasma control. III. Genetic analysis of chiasma frequency variation in two selected lines of *Schistocerca gregaria* Forsk. *Chromosoma* **46:** 365–374.

Shepher, J. (1971). Mate selection among second generation kibbutz adolescents and adults: incest avoidance and negative imprinting. *Arch. sex. Behav.* **1:** 293–307.

Slatkin. M. (1975). Gene flow and selection in a two-locus system. *Genetics* **81:** 787–802.

Snyder, R.L. (1976). *The Biology of Population Growth.* London: Croom Helm.

Stebbins, G.L. (1950). *Variation and Evolution in Plants.* Columbia University Press.

Stebbins, G.L. (1971). *Chromosomal Evolution in Higher Plants.* London: Edward Arnold.

Street, H.E. (1974). *Tissue Culture and Plant Science.* New York: Academic Press.

Strobeck, C. (1975). Selection in a fine-grained environment. *Am. Natur.* **109:** 419–425.

Strobeck, C., Maynard Smith, J. & Charlesworth, B. (1976). The effects of hitch-hiking on a gene for recombination. *Genetics* **82:** 547–558.

Suomalainen, E. (1950). Parthenogenesis in animals. *Adv. Genet.* **3:** 199–253.

Suomalainen, E. & Saura, A. (1973). Genetic polymorphism and evolution in parthenogenetic animals. I. Polyploid Circulionidae. *Genetics* **74:** 489–508.

Sved, J.A. (1968). Possible rates of gene substitution in evolution. *Am. Natur.* **102:** 283–293.

Sved, J.A., Reed, T.E. & Bodmer, W.F. (1967). The number of balanced polymorphisms that can be maintained in a natural population. *Genetics* **55:** 469–481.

Talmon, Y. (1964). Mate selection in collective settlements. *Am. sociol. Rev.* **29:** 491–508.

Teitelbaum, M.S. (1972). Factors associated with the sex ratio in human populations. In *The Structure of Human Populations,* eds. G.A. Harrison & A.J. Boyce, pp. 90–109. Oxford University Press.

Thomson, G. (1977). The effect of a selected locus on linked neutral loci. *Genetics* **85:** 753–788.

Thomson, G., Bodmer, W.F. & Bodmer, J. (1976). The HL-A system as a model for studying the interaction between selection, migration, and linkage. In *Population Genetics and Ecology,* ed. S. Karlin & E. Nevo, pp. 465–498. New York: Academic Press.

Tomlinson, J. (1966). The advantage of hermaphroditism and parthenogenesis. *J. theoret. Biol.* **11:** 54–58.

Trivers, R.L. (1972). Parental investment and sexual selection. In *Sexual Selection and the Descent of Man,* ed. B. Campbell, pp. 136–179. Chicago: Aldine.

Trivers, R.L. (1974). Parent–offspring conflict. *Am. Zool.* **14:** 249–265.

Trivers, R.L. (1976). Sexual selection and resource – accruing abilities in *Anolis garmani. Evolution* **30:** 253–269.

Trivers, R.L. & Hare, H. (1976). Haplodiploidy and the evolution of the social insects. *Science* **191:** 249–263.

Trivers, R.L. & Willard, D.E. (1973). Natural selection of parental ability to vary the sex ratio of offspring. *Science* **179:** 90–92.

Turner, J.R. (1967). Why does the genome not congeal? *Evolution* **21:** 645–656.

Uzell, T.M. & Goldblatt, S.M. (1967). Serum proteins of salamanders of the *Ambystoma jeffersonianum* complex, and the origin of triploid species of this group. *Evolution* **21:** 345–354.

Van Valen, L. (1973). A new evolutionary law. *Evol. Theory* **1:** 1–30.

Vanzolini, P.E. (1970). Unisexual *Cnemidophorus lemniscatus* in the Amazonas valley: a preliminary note (Sauria, Teiidae). *Pap. avulsos Dep. Zool., Sao Paulo* **23:** 63–68.

Verner, J. & Willson, M.F. (1966). The influence of habitats on mating systems of North American birds. *Ecology* **47:** 143–147.

Verner, J. & Willson, M.F. (1969). Mating systems, sexual dimorphism, and the role of male North American passerine birds. *Ornithol. Monogr.* **9:** 1–76.

Von Neumann, J. & Morgenstern, O. (1953). *Theory of Games and Economic Behaviour*. Princeton University Press.

Wagoner, D.E., McDonald, I.C. & Childress, D. (1974). The present status of genetic control mechanisms in the house fly, *Musca domestica* L. In *The Use of Genetics in Insect Control*, ed. R. Pal & M.J. Whitten, pp. 183–197. Amsterdam: Elsevier.

Wallace, B. (1968). Polymorphism, population size and genetic load. In *Population Biology and Evolution*, ed. R.C. Lewontin, pp. 87–108. Syracuse University Press.

Weir, J.A. (1953). Association of blood-pH with sex ratio in mice. *J. Hered.* **44:** 133–138.

Weir, J.A. (1962). Hereditary and environmental influences on the sex ratio of PHH and PHL mice. *Genetics* **47:** 881–897.

Weir, J.A. & Clark, R.D. (1955). Production of high and low blood-pH lines of mice by selection with inbreeding. *J. Hered.* **46:** 125–132.

Werner, P.A. & Platt, W.J. (1976). Ecological relationships of co-occurring goldenrods (*Solidago*: Compositae). *Am. Natur.* **110:** 959–971.

Westergaard, M. (1958). The mechanism of sex determination in dioecious flowering plants. *Adv. Genet.* **9:** 217–281.

White, M.J.D. (1966). Further studies on the cytology and distribution of the Australian parthenogenetic grasshopper, *Moraba virgo*. *Revue suisse Zool.* **73:** 383–398.

White, M.J.D. (1973). *Animal Cytology and Evolution*, 3rd edition. Cambridge University Press.

Wiley, R.H. (1973). Territoriality and non-random mating in sage grouse, *Centrocercus urophasianus*. *Anim. Behav. Monog.* **6**(2): 85–169.

Wiley, R.H. (1974). Evolution of social organisation and life-history patterns among grouse. *Q. Rev. Biol.* **49:** 201–227.

Williams, G.C. (1966). *Adaptation and Natural Selection*. Princeton University Press.

Williams, G.C. (1975). *Sex and Evolution*. Princeton University Press.

Williams, G.C. & Mitton, J.B. (1973). Why reproduce sexually? *J. theoret. Biol.* **39:** 545–554.

Winterbourne, M. (1970). The New Zealand species of *Potamopyrgus* (Gastropoda: Hydrobiidae). *Malaco logia,* **10:** 283–321.

Wood, R.J. & Newton, M.E. (1976). Meiotic drive and sex ratio distortion in the mosquito *Aedes aegypti*. *Proc. int. Congr. Ent.* pp. 97–105.

Wynne-Edwards, V.C. (1962). *Animal Dispersion in Relation to Social Behaviour*. Edinburgh: Oliver and Boyd.

Young, P. (1976). Ph.D. thesis, University of Cambridge.

Zaffagnini, F. & Sabelli, B. (1972). Karyological observations on the maturation of the summer and winter eggs of *Daphnia pulex* and *Daphnia middendorffiana*. *Chromosoma* **36:** 193–203.

Zahavi, A. (1975). Mate selection – a selection for a handicap. *J. theoret. Biol.* **53**: 205–214.

Zahavi, A. (1977). The cost of honesty (further remarks on the handicap principle). *J. theoret. Biol.* **67**: 603–605.

Zaret, T.M. (1972). Predator–prey interaction in a tropical lacustrine system. *Ecology* **53**: 248–257.

Author index

Abdullah, N.F. 74, 75
Acton, A.B. 74
Adams, M.S. 143
Alexander, R.D. 144, 166
Allard, R.W. 74, 75, 128, 129, 130
Amadon, D. 185
Assem, J. van den 162

Babcock, E.B. 65
Bacci, G. 150
Baker, H.G. 99, 109, 135, 136, 137
Baker, R.R. 151
Banta, A.M. 59
Bar-Anan, R. 150
Basrur, V.R. 56
Bateman, A.J. 169
Bawa, K.S. 136, 137
Bell, G. 151
Bellrose, F.C. 181
Bengtsson, B.O. 139
Bergerard, J. 63
Birky, C.W. 61
Bodmer, W.F. 19, 24, 78, 86
Breder, C.M. 182, 183
Brues, M.A. 24
Bruning, D.F. 182
Bull, J.J. 130, 133, 163
Bulmer, M.G. 141

Carlisle, T.R. 175
Carlquist, S. 136
Carson, H.L. 43
Catcheside, D.G. 77
Charlesworth, B. 30, 74, 75, 84, 87, 89, 95, 120, 121, 158
Charlesworth, D. 84, 87, 120
Charnov, E.L. 130, 131, 133, 162, 163, 166
Chewyreuv, I. 162
Childress, D. 164
Chinnici, J.P. 74, 75, 76
Christensen, B. 45
Clark, R.D. 149

Clarke, B. 25
Clarke, C.A. 85
Clayton, G.A. 11
Clegg, M.T. 128
Clutton-Brock, T.H. 179, 184, 186
Collier, G. 181
Cox, E.C. 112, 113, 190, 191, 192
Craig, G.P. 164
Crosby, J.L. 86
Crow, J.F. 19, 21, 24
Cuellar, O. 46, 47, 48, 56, 67

Darlington, C.D. 23
Darwin, C. 42, 136, 168, 171, 183, 184
Davis, G.W.F. 173
Dawkins, R. 113, 175
Detlefson, J.A. 74, 75
Dewees, A.A. 74, 75
Dobzhansky, Th. 184
Drake, J.W. 188, 189, 190
Drickamer, L.C. 142

Edwards, A.W.F. 148, 149
Ehrman, L. 171
Emerson, A.E. 2
Ernst, A. 86
Eshel, I. 14, 82

Falconer, D.S. 76, 149, 170
Feldman, M.W. 14, 78, 82
Felsenstein, J. 10, 15, 22, 32, 33, 73, 78, 117, 121, 122, 123
Fisher, R.A. 2, 3, 13, 17, 32, 40, 77, 117, 119, 129, 146, 147, 157, 159, 160, 171, 172, 173, 174
Fossey, D. 142
Frame, G.W. 152
Frame, L.H. 142
Franklin, I. 78, 82
Freeman, D.C. 162
Frith, H.J. 182

Ghiselin, M.T. 133, 135, 138

Gibson, T.C. 112, 113, 192
Gilbert, J.L. 61
Gilpin, M.E. 2
Glesener, R.R. 109, 110
Godley, E.J. 136
Goldblatt, S.M. 47
Grant, V. 109
Greene, P. 144
Greenwood, P.J. 148

Haas, R. 172
Haigh, J. 30, 34, 111, 112
Haldane, J.B.S. 23, 24
Hamilton, W.D. 114, 146, 147, 160, 161, 164, 165
Harcourt, A.H. 142
Hare, H. 165, 166
Harper, J.L. 156
Harper, K.T. 162
Harrison, B.S. 12
Harvey, P.H. 141, 179, 184, 186
Haskins, C.P. 173
Hebert, P.D.N. 59
Hickey, W.A. 164
Hill, J.L. 140
Hill, W.G. 10, 32, 111, 122, 123
Howe, H.F. 159
Hubby, J.L. 24
Hutchinson, G.E. 60

Jain, S.K. 130
Jenni, D.A. 181, 184
Jinks, J.L. 129
John, B. 120

Kahler, A.L. 128
Karlin, S. 15, 78, 82
Kaufman, D.W. 130
Kear, J. 180
Kerfoot, W.C. 60
Kidwell, M.G. 74, 75, 76
Kimball, A.W. 149, 159
Kimura, M. 19, 21, 24, 78, 188, 189, 194
King, J.L. 24
Klikoff, L.G. 162
Knowlton, N. 155
Knox, P.B. 62
Kojima, K. 78, 82

Lack, D. 179
Langley, C.H. 84
Lee, L. 61
Leigh, E.G. 188, 189, 194

Lerner, I.M. 129
Levene, H. 92
Levin, D.A. 109, 110
Levins, R. 131
Levi-Strauss, C. 143
Lewis, D. 147
Lewis, K.R. 120
Lewontin, R.C. 24, 78, 81, 82, 84
Lokki, J. 57
Lovell, P.H. 156

McAlister, W.H. 156
Macarthur, R.H. 27, 147, 167
McDonald, I.C. 164
Marquiss, M. 159
Maslin, P.T. 46
Mather, K. 12, 81, 129
Maynard Smith, J. 2, 3, 14, 15, 17, 20, 22, 24, 26, 28, 29, 30, 31, 90, 93, 96, 101, 104, 106, 111, 112, 126, 127, 130, 132, 147, 148, 168, 173, 174, 175, 176
Milkman, R.D. 24
Mitton, J.B. 91, 99
Mockford, E.L. 53, 54
Moore, K.G. 156
Moore, W.S. 69
Moran, P.A. 17
Morgenstern, O. 147
Mukai, T. 85, 189
Muller, H.J. 4, 18, 32, 33, 36, 117
Muntzing, A. 65, 66
Murray, J. 86, 135

Neaves, W.B. 46
Neel, J.V. 142, 143
Newton, I. 159
Newton, M.E. 164
Nikolskii, G.V. 156
Novitski, E. 149, 159
Nur, U. 43

O'Donald, P. 172, 173, 184
Ohta, T. 19
Olsen, M.W. 45
Opler, P.A. 136, 137
Orians, P.H. 180

Packer, C. 141, 142
Parker, E.D. 57
Parker, G.A. 151, 154, 155, 178
Pijnacker, L.P. 44, 63
Pimentel, R.A. 156
Platt, W.J. 156

Poulsen, D.F. 147
Preston, F.W. 27
Price, G.R. 147, 176
Pusey, A. 142

Ralls, K. 185
Reed, T.E. 24
Rees, H. 120
Rensch, B. 186
Richards, A.J. 55
Roberts, E. 74, 75
Robertson, A. 10, 11, 32, 111, 122, 123, 150
Robertson, J.G. 64
Rosen, D.E. 182, 183
Rothfels, K.H. 56
Ruttner-Kulisko, A. 61

Sabelli, B. 59
Sakaguki, B. 147
Saura, A. 57
Schull, W.J. 142
Schultz, R.J. 47
Scudo, F.M. 149
Seiler, J. 45
Selander, R.K. 57, 130, 171, 185
Shaw, D.D. 74, 75, 76
Shepher, J. 143
Sheppard, P.M. 85
Sherman, P.W. 166
Slatkin, M. 97, 98
Smith, V.G.F. 151
Snyder, R.L. 163
Stebbins, G.L. 23, 45, 49, 51, 54, 55, 65, 109
Stewart, K.S. 142
Street, H.E. 51
Strobeck, C. 30, 91, 111, 116, 120, 121
Suomalainen, E. 54, 57, 109
Sved, J.A. 24

Talmon, Y. 143
Teitelbaum, M.S. 142, 159
Thomson, G. 30, 78, 86, 112
Tilman, D. 109, 110
Tomlinson, J. 134, 135
Trivers, R.L. 133, 146, 147, 159, 162, 165, 166, 174, 175, 178, 184
Turner, J.R. 78, 117

Uzell, T.M. 47

Van Valen, L. 26, 54, 110
Vanzolini, P.E. 48
Verner, J. 180, 184
Vessey, S.H. 142
Von Neumann, J. 147

Wagoner, D.E. 164
Wallace, B. 24
Weir, J.A. 149, 150
Werner, P.A. 156
Westergaard, M. 163
White, M.J.D. 23, 43, 45, 46, 48, 49, 54, 56, 63
Wiley, R.H. 171, 180, 184
Willard, D.E. 133, 147, 159, 162
Williams, G.C. 2, 5, 9, 10, 52, 57, 63, 66, 69, 89, 90, 99, 107, 113
Willson, M.F. 180, 184
Wilson, E.O. 27
Winterbourne, M. 64
Wood, R.J. 164
Wynne-Edwards, G.C. 2

Yamaguchi, O. 85
Yokoyama, S. 10, 111, 117, 121, 122, 123
Young, P. 59, 60

Zaffagnini, F. 59
Zahavi, A. 173, 174
Zaret, T.M. 60

Subject index

Acarini, arrhenotokous families and sub-families of, 53, 160

Acarophenax (mite): viviparous, with single male in litter mating with females before birth, 160

Accipiter nisus (sparrowhawk), sexual dimorphism and sex ratio in, 159

Acer negundo (box elder): sex ratio of, in dry and damp sites, 162

Actitis macularis (spotted sandpiper), polyandry in, 181

Aedes aegypti (mosquito): sex ratio in, 164

Aleurodidae (white flies), arrhenotoky in, 53

Amblystoma (salamander): thelytokous forms of, 47, 48

Anas (river ducks), uniparental care in, 181

anisogamy, evolution from primitive isogamy to, 39, 146, 151–5

Anolis garmani (lizard), large males mate more often, 184

Anser (geese), biparental care in, 181

Antennaria (Compositae), apomictic species complex of, 55

aphids, cyclical parthenogenesis in, 62

apomixis, in thelytokous parthenogenesis, with suppression of meiosis, 43; in animals, 45–6; application of term to plants, 49; facultative, in plants, 55, 56; in plants, commonly associated with polyploidy, 50–1, 55; possible origin of, 48

arrhenotoky, production of males from unfertilised eggs, 42; combined with inbreeding and excess of females in mites and insects, 160; taxonomical distribution of, 54

asexual reproduction, in plants: by bulbs, runners, etc., 49; by seeds produced asexually, 50

Atriplex confertifolia (shadscale): sex ratio of, in dry and damp sites, 162

automixis, in thelytokous parthenogenesis, 43–5, 63; artificial selection for, 47; in plants, 50; with pre-meiotic doubling, in Moraba, 46, 56

Avena barbata, A. fatua: partial outcrossing in, 128, 130

Avena sativa, plants produced by outcrossing have higher fitness, 129–30

Aythya (diving ducks), uniparental care in, 181

Bacillus rossius (stick insect): bisexual in part of range, automictic elsewhere, 44, 63

bacteriophages lambda and T4, mutation rates in, 189

'balance' argument: for maintenance of cyclical parthenogenesis, 59–60, 62; for maintenance of recombination, 9, 72; for short-term advantages of sex, 5–6, 38, 57–8, 66–7

Biomphalaria (snail): self-compatible, but outcrosses if opportunity occurs, 134

birds: avoidance of inbreeding in, 140–1; parental care in, 179–83; size dimorphism in, 185

blood pH: artificial selection for, in mice, with accidental association with sex ratio, 149–50

Bonasa umbellus (ruffed grouse), male with central territory obtains most matings, 170

Bosmina longirostris (Cladocera): seasonal dimorphism in, related to predation, 60

Bromus mollis (grass), partial outcrossing in, 128

brooding of young: association of hermaphroditism with, 138; resource allocation and, 139

Calidris alba (sanderling), parental care in, 181

Calligrapha spp. (chrysomelid beetles), facultative parthenogenesis in, 64, 67

Caryophyllaceae, dioecy in, 135–6

cattle, sex ratio in, 150

Cepaea (snail), self-incompatible hermaphrodite, 125

Ceriodaphnia cornuta (Cladocera), dimorphism of, 60

chiasma formation: genetic variation in frequency and location of, 9, 74; in scheme for origin of sex, 7, 8; selectively favoured? 72–3; *see also* recombination

chimpanzee, female transfer in, 142

choice, female, 168–71; of 'handicaps' in male, 171–4; self-reinforcing of, 171–2, 173–4

chromosomes: in scheme for origin of sex, 8; of *Aedes aegypti*, 164

Clitumnus extradentatus (stick insect), automictic, 63

Cnemidophorus uniparens (whiptail lizard), parthenogenetic, 46, 48, 56

Cnemidophorus tesselatus, parthenogenetic: diploid and triploid biotypes of, 57

Cnephia mutata (simuliid fly), triploid parthenogenetic biotypes, 56

Collinsia sparsiflora (Scrophulariaceae), varies from complete selfing to random mating, 128

colonisation of new region: parthenogenesis in, 110; by single hermaphrodite individuals, 134, 136, 139; by two populations, and selection for recombination, 15, 96–7

competition: between *E. coli* wild-type and mutator strains (non-recombinant), 112, 113; between sexual and asexual populations in fluctuating environment, 92–5; for mates, and sex ratio, 160–1

Crepis (Compositae), apomictic species complex of, 55

cyclomorphosis, in Cladocera, 60, 61

Cygnus (swans), biparental care in, 181

Cynipidae (gall wasps), parthenogenesis in, 61

cytoplasmic 'genes', 147

Daphnia (Cladocera): cyclical apomixis in, 58–60, 67, 70

Daphnia pulex, D. middendorffiana: all-female populations of, producing 'winter eggs' parthenogenetically, 59

Dichanthium aristatum (grass): balance between sex and parthenogenesis in, 62–3, 67, 70

Dinophilus, sex determination before fertilisation in, 150

dioecy, 125; in animals, 133; in plants, 133, 136, 138; and avoidance of inbreeding, 135

diploidy, possible advantages of, 9

Diptera: chiasmata absent in meiosis in males of, 9, 73; 'rare male' effect in, 140, 171; thelytokous, 56

dispersion pattern: and avoidance of inbreeding, 140–1; and effect of sib competition on selection for recombination, 108; seeds and, in plants, 49

Distichlis spicata (salt grass): sex ratio of, in dry and damp sites, 162

DNA: recombination and repair of, 7, 193

Drosophila mangabeirai, automixis in, 44–5

Drosophila melanogaster: linkage disequilibrium in, 84–5; mutation rate in, 189–90; selection for increased or decreased recombination frequency in, 74, 75, 76

Drosophila mercatorum, automixis in, 43

Drosophila spp.: inversion polymorphisms in, 88; no evidence for genetic variance of sex ratio in, 149; non-random association between alleles determining isozymes and chromosome order in, 87; 'sex ratio' condition in, 147

Drosophila subobscura: female choice in, 168, 169–70, 175; inbreeding and survival of offspring of, 126, 127, 129

dry habitats: favour parthenogenesis, 109, and high rates of selfing in hermaphrodites, 128; sex ratio of dioecious plants in, 162–3

earthworms, parthenogenetic, 46

ecosystems: convergent and divergent, 27–8; number of species in, as equilibrium between speciation and extinction, 27

eggs: cleavage of, initiated by sperm without contribution of genetic material (*amblystoma*), 47; development of unfertilised, not uncommon, 43, 49; size of mother and size of, 156; winter, of *Daphnia*, 58–9, and rotifers, 6, 61

enchytraeid oligochaets, thelytokous, 45

endomitosis (pre-meiotic doubling of chromosomes), in some examples of thelytoky, 46, 48

environment: evolutionary advance by one species experienced by other species as deterioration of, 26, 110; how rapidly can it change without causing extinction? 21–2; how much more rapidly can a sexual population adapt to, than an asexual one? 22; parthenogenesis tends to occur in biologically simple, 109–10; and sex ratio, 162–3; spatially varying, with migration, selection for sex and recombination in? 89, 96–9; variable or unpredictable, selection for sex and recombination in? 4, 10, 89, 90–6, (evidence from geographical distribution of parthenogenesis) 108–9

enzymes, polymorphic, 24; frequencies of, as measure of frequency of selfing in plant populations, 128; identification of biotypes by study of, 57

Ephedra viridis (mormontea): sex ratio of, in dry and damp sites, 162

Escherichia coli: competition between wild-type and mutator strains of (non-recombinant), 112, 113; mutation rate in, 189, 190–1

eukaryotes: origin of, 6; scheme for origin of meiosis in, 7, 8

evolution: potential of parthenogenetic strains for, 51–7; transient and steady-state models of, in asexual population, 30–1, and in sexual population, 28–30

evolution rate: compared with rate of change in artificial selection experiments, 11–12; depends on balance between exhaustion of variability by directional selection, and its generation by mutation, 12–13; limit on (cost of selection), 23, 24; mutation rate and, 30; in response to continuous change of environment, 19–23; in response to selection based on pre-existing genetic variability, 13–16; in response to sudden change in environ-

ment, 16–19; sex and recombination and, 11, 32

evolutionarily stable strategy (ESS), 147–8, 172; parental care in terms of, 175, 176–9

extinction of species: balance between parthenogenetic mutations and, 70; lack of evolutionary potential and, 52, 54; law of constant (Van Valen), 26–7; relation between evolution rate, lag load, and, 19–23; simulation of random process of speciation and, 52

female choice, 168–71; in maintenance of sexual reproduction? 174; of males with 'handicaps', 171–4; and sexual dimorphism, 183–4

females: natural populations with excess of, 6, 62, 64, 65, 160; species with larger, 185

fertilisation: internal or external, and male parental care, 178–9

fish (bony): parental care in, 182–3; parthenogenetic species of, 47

fitness, genetic: estimate of, for parthenogenetic and sexual forms of *Poeciliopsis*, 68–9; 'inclusive', 114, 165; low, of inbred offspring, 124; of males with 'handicaps', 173; mean, of a population, 21, 122–3; additive genetic variance in, 170, 171, 172; variance of, between sexes, and sex ratio, 162–3

fitness sets: for allocation of resources in hermaphrodites, 130–2; factors favouring convex set, in plants, 132–3

flukes, digenetic, 61

forest plants: proportions of hermaphrodite, monoecious, and diœcious among, 136, 137

frogs, parental care in, 182

games, theory of, 147, 175, 176

genes: number of, and mutation rate, 191; for recombination frequency, 76; 'hitch-hiking' and selection of, 112–14; ways in which sex ratio could be influenced by, 146–7, 157

gonochoristic, use of term, 125

gorilla, female transfer in, 142

group selection: argument for maintenance of sex by, 2–5; argument weaker for plants, 51; 'balance' argument against, 5–6; and coexistence of

sexual and parthenogenetic forms, 68; and cyclical parthenogenesis, 59, 62; fails to explain origin of sex, or maintenance of recombination, 6, 9, 71; share of, in maintaining sex, 6, 37–8, 69–71
grouse, parental care in, 180
gynodioecious plants, 42, 147
gynogenesis (females produce eggs parthenogenetically after mating), 47, 50

haplo-diploid cycle, scheme for origin of, 7, 8–9
haplo-diploid systems, 42, 43; sex ratio in, 150
Hawaiian Islands, dioecy in flora of, 136
Helix, self-incompatible hermaphrodite, 125
Hemidactylus (geckos), parthenogenesis in, 46
Hermaphroditism, 125; advantages of, 39–41, 50; in animals, 138–9; models for evolution of, 130–5; as primitive condition in higher plants, 135; resource allocation in, 40, 130–3; selection for self-compatibility in, 125–6; selection for self-sterility in, 126–7; self-sterility in, partial, 127–8; sequential, 135
heterokaryon: evolution to diploid cell from, 7, 8; selective advantage of, 7
heterostyly, 86, 125, 136
Hieracium (Compositae), apomictic species complex of, 55
Hill–Robertson effect, 10, 32, 122
histones: one amino-acid difference in, between animals and plants, 30
'hitch-hiking' effect, 10; and gene selection, 112–14; in models for evolution, 30; and rate of fixation of mutations, 112, 113, 192–3; and recombination, 114–17, 123
Hordeum vulgare, partial outcrossing in, 128
horse ancestors: rate of change of molar teeth of, 11–12
Houttuynia cordata (Saururaceae), apomict with no sexual relatives, 31, 65
humans: incest taboo in, 142–5; sex ratio in, 158–9, (no evidence for genetic variance of) 148–9, (reproductive condition of mother and) 162

hybrid vigour, 7, 9; less marked in habitually selfing species? 129
hybridisation: and origin of sexual dimorphism, 184; association with parthenogenesis, 48, 50–1
Hymenoptera: arrhenotoky and thelytoky in, 42–3, 52–3; sex ratio in, 150, 165–6, (size of host and, in parasitic species), 162

Iceryini (Homoptera), arrhenotoky in, 53
inbreeding: in arrhenotokous species of insects and mites, with large excess of females, 160; avoidance of, in animals, 139–42, and in humans, 142–5; avoided in plants by dioecy and monoecy, 135; and chiasma formation in rye, 120; and fitness of offspring, 124, genetic consequences of protracted, 129–30; in hermaphrodite plants, 160; mechanisms to prevent, 125; and selection for higher recombination, 119–22; and sex ratio, 160–1
incest taboo, in humans, 142–5
inversions: and linkage disequilibrium, 87–8; non-random association of, with enzyme loci 84; in thelytokous diptera, 56
islands: colonists of, 136; number of species on, 27
isogamy, 38–9; evolution to anisogamy from, 39, 146, 151–5; no advantage of parthenogenesis in, 42

Jacana spinosa, polyandry in, 181

kibbutzim, Israel: marriages of children communally reared in, 143

Lacerta (wall lizards), parthenogenesis in, 46
lag load, *see* load, genetic
Lagopus lagopus (red grouse), parental care in, 150
Leipoa ocellata (mallee fowl), parental care in, 182
'lek' species of birds, 170–1
lima bean, selection for recombination frequency in, 74, 75
linkage disequilibrium: in natural populations, 73, 83–5, 85–8; and selection for reduced recombination, 80–1; processes acting to maintain, 81–3,

randomly generated, favours recombination, 112, 122, 123; and rate of evolution, 14–16

lizards, parthenogenetic species of, 46

load, genetic: lag load, 21–2, 26; concept of, irrelevant for intraspecific competition, 26; criticisms of concept of, 23–5; upper critical value of, at which population becomes extinct, 22

Lolium multiflorum (grass), partial outcrossing in, 128–9

Lycaon pictus (hunting dog), female transfer in, 142

Macaca mulatta, male transfer in, 142

male sterility: in gynodioecious plants, 147; selection for, in parthenogenetic varieties derived from hermaphrodite ancestors, 41–2

males: female choice of, 140, 171–4; parental care by, 39, 42 (*see also* parental care)

mammals: with larger females, 185; parental care in, 179

Marmota monax (woodchuck), sex ratio in, 163

mates: hermaphrodites' advantage in finding, 134; local competition for, and sex ratio, 160–1

meiosis: elimination of male genes during, in *Poeciliopsis* 47; in eukaryotic sex, 8, 9, 39; in male Diptera (chiasmata absent), 9, 73; and sex ratio, 160; suppressed in apomixis, 43, 45; usually prerequisite for development in animals, not in plants, 51, 70

meiotic drive, 147, 164

mice, sex ratio in, 149–50

Micromalthus debilis (beetle), arrhenotoky in, 53

migration: into islands, 27, 136; and selection for recombination, 13, 96–9

monoecy, 125, 135; ensures outcrossing, 138; in forest trees, 137, 138

monogamy: biparental care common in, 183; in carnivores and primates, 179

Moraba virgo (grasshopper), automixis in, 46, 56

'Muller's ratchet', 10, 33–6, 119, 122; and mutation rate, 191, 192

mutations: balance between selection and, 12–13; chain-terminating, 190; diploidy as protection against deleter-

ious, 9; fixation of, 17–19, 28; 'hitch-hiking' and rate of fixation of, 30, 112, 113, 192–3; rate of, 188, 192; rate of, in infinite asexual population, 193–4; rate of, and rate of evolution, 30; selection for recombination in presence of, 117–19

mutator gene in *E. coli,* 112, 113, 192–3

Neomorpha acutirostris (huia), sex differences in, 185

Neurospora crassa: mutation rate in, 109; mutations in recombination rate of, 77

Notobranchius guentheri (cyprinodont fish), 172–3

Otiorrhynchus scaber (weevil), parthenogenesis in, 57

Oxyura (stifftail ducks), uniparental care in, 181

Papilio memnon (butterfly), supergene controlling mimicry patterns in, 85–6, 87

Papio anubis (baboon), male transfer in, 141

parental care, 175, (paternal) 39, 42; models of strategies for, 176–9, (applied to vertebrates) 179–83

parthenogenesis: in animals, 5, 42–9; arrhenotoky and thelytoky in, 42–3; coexistence of genetically different clones in, 31; coexistence of sexual species and, 58, 66–9; cyclical, 58–62, 70; evolutionary potential in, 5, 51–7; facultative, 5, 62–5, 70; geographical distribution of, 26, 57, 108–10; group selection and, 5, 27–8, 68; meiotic, with pre-meiotic doubling of chromosomes, 46, 56; origins of, 47–9; in plants, 49–51; pseudogamy in, 50, 65–6; short-term balance between sexual reproduction and, 62–3; twofold advantage of, 2–4, (factors invalidating advantage) 38–42

Parus major (great tit), inbreeding in, 141

Passer domesticus (house sparrow), dimorphism in, 185

peacocks, female choice in, 171, 173

Peromyscus maniculatus (deer mouse), inbreeding avoidance in, 140

Phasmidae (stick insects), facultative parthenogenesis in, 63 planktonic animals, hermaphroditism in, 138, 139

Poa (grass): apomictic species complex of, 55, 70; pseudogamy in, 65

Poecilia formosa, parthenogenesis in, 47, 48, 50

Poeciliopsis, parthenogenesis in, 47, 48, 68–9

pollen, fertilisation of endosperm by (pseudogamy), 50, 65–6

pollination: by insects, in plants with hermaphrodite flowers, 132; by wind or water, in dioecious and monoecious plants, 133, 136, 138

polyandry, in birds, 181

polygamy: sexual selection more intense in, 184; size of males, and mating success in, 133

polygyny: in birds, 180–1; and sexual dimorphism, 186–7

polyploidy: apomixis associated with, in plants, 50–1, 55; in thelytokous enchytraeids, 45

Potamopyrgus antipodorum (snail), facultative parthenogenesis in, 64–5, 67

Potentilla (Rosaceae): apomictic species complex of, 55, 66, 67, 70; pseudogamy in, 65

primates: female transfer in, 142; monogamy in, 179; sexual dimorphism in, as function of body size, 186

Primula, supergene controlling heterostyly in, 86

prokaryotes: capacity for genetic recombination in, 7; mutation rates in, 190

protandry in hermaphrodites: 134; in plants, 125

protogyny in hermaphrodites, 134

pseudogamy, 50, 65–6

psocids, thelytoky among, 53, 54

Quiscalus quiscula (grackle), sex ratio in, 159, 171

recombination, genetic: as adaptation to biologically complex environment, 110; cost of, 85; delays accumulation of harmful alleles, acting as a form of repair, 33–6; and evolution rate, 11, 32; genes affecting, may act locally or generally, 76; genes for high rate of, 112–13; genetic variance for frequency of, 72, 73–5, (nature of variance) 76–7;

group selection unable to explain maintenance, of, 6, 9, 71; 'hitch-hiking' and, 114–17, 123; origin of, in repair processes for prokaryote DNA, 6, 7; selection against, at equilibrium in uniform environment, 72, 77–83, 89; selection for, in a fluctuating environment, 89, 95–6, in a new region, 15, 96, in presence of recurrent mutations, 117–19, and in spatially varying environment, 89, 96–9; short-term advantages of, 10; short-term disadvantages of, 9, 73; selfing, and selection for higher rate of, 112, 119–22; *see also* sexual reproduction

red deer, fighting of males of, 169

Red Queen hypothesis, 26, 54, 110

resource allocation: to brooding of young, 139; to male and female functions in hermaphrodites, 40, 130–3

Rhea americana, parental care in, 182

rotifers: bdelloid, without males, 53, 69; of order Monogononta, 53, 60; with overwintering sexual and immediately hatching asexual eggs, 6, 61

Rubus subgenus *Eubatus* (blackberries), apomictic species complex of, 55

Rumina (snail), self-compatible hermaphrodite, with occasional outcrossing, 130

rye: chiasma frequency in, 120

Salmonella typhimurium, mutation rate in, 187

Schistocerca gregaria, selection for recombination frequency in, 74, 75, 76

seeds, size of, 156

selection: artificial, for automixis, 47, for blood pH in mice, 149–50, for number of abdominal bristles in flies, 11, 12, and for recombination frequency, 74, 75; cost of (Haldane), 23, 24; group and individual, 1–2 (*see also* group selection); 'hard' and 'soft', 24–5, 25–6; kin, 66; models for evolution with intermittent, 28–9, 30, and with long-continued, 29, 30; 'second order', on genes altering mutation and recombination rates, 112–13; sexual, 10, 168–87; threshold, 24, 25

self-fertility: advantageous if a mate is not available, 124; selection for, in hermaphrodites, 125–6

self-sterility, in plants, 125, 126–8
selfing species: selection for self-incompatibility in, 125–9; in tropical environments, 26; selection for higher recombination in, 112, 119–22
sessile habit: association of hermaphroditism with, in animals, 138, 139; diminishing returns from increased production of seeds and pollen in, 133
sex, mechanisms for determining, 164
sex ratio, 40, 64, 146–8; genetic variance in, 148–50; with local competition for mates, 160–1; with random mating, 157–60; in social Hymenoptera, 165–6; stable, 157, 166–7; when value of male and female varies with circumstances, 161–3
sexual dimorphism: barrier to hybridisation in early stages of speciation, and production of, 184; ecological differentiation and, 184–5; polygamy and, 184, 186; sexual selection in production of, 183–4; in size, as function of body size in primates, 186; in size, with larger females, 185
sexual reproduction: as adaptation to biologically complex environment, 90, 109–10; and evolution rate, 11, 32; female choice in maintenance of? 174; long-term advantages of, 4, 37; is it maintained by group selection? 37–8; origin of, 6, 7–9, 71; and rate of evolution, 22–3; share of group selection in maintaining, 6, 69–70, 70–1; short-term advantage of, 5, ('balance' argument for) 5–6, 38, 57–8, 66–7, (with sib competition) 89–90, 99–108, (in fluctuating environment?) 4, 10, 89, 90–6, (in spatially varying environment?) 89, 96–9; short-term balance between parthenogenesis and, 62–3; *see also* recombination, genetic
sexual selection, 10, 168; female choice in, 168–74; monogamy, polygamy, and sexual dimorphism in, 183–7; parental care and, 175–83
sib competition: between offspring of sessile organisms, 133; effect on, of many loci concerned with a single environmental feature, 107–8; patterns of dispersal and, 108; short-term advantages for sex and recombination, 10, 99–100, 123; simulation of, 100–7

Sipyloidea sipylus (stick insect): parthenogenesis in, 63, 67
skin grafts, showing genetic identity of different populations of *Cnemidophorus*, 56
skua, Arctic: sexual selection in, 184
snails, supergenes controlling shell colour and pattern in, 86
Solenobia triquetrella (moth), parthenogenesis in, 57
Solidago (golden rod): size of seeds in, 156
species abundance distribution (Preston), 27
Spergula arvensis (corn spurrey), genetic polymorphism of seeds of, 156
stickleback, polygyny in, 183
supergenes, promote linkage disequilibrium, 85–7
syngamy, in eukaryotic sex, 8, 9, 39

Taraxacum (dandelions): asexual clones of, produce functionless pollen, 41; species complex of, 54–5, 67
Thalictrum fendleri (meadow rue): sex ratio of, in dry and damp sites, 162, 163
thelytoky (production of females from unfertilised eggs), 43; geographical distribution of, 54; with normal meiosis and subsequent fusion of pronuclei or early cleavage nuclei (automixis), 43–5, 47–8; with pre-meiotic doubling of chromosomes, 46, 48; with suppression of meiosis (apomixis), 43, 48, (in animals) 45–6, (in plants) 49, 50–1, 54, 55; taxonomic distribution of, 53, 54
Thysanoptera (thrips), arrhenotoky in, 53
trees: hermaphrodite, monoecious, and dioecious, 136, 137; pollination of, 136, 138
Tribolium castaneum, selection for recombination frequency in, 74, 75
tropics: forest trees in, 136, 137; parthenogenesis rare in, 109
turkeys, hatching of unfertilised eggs of, 45

variability, genetic: generation of, by chiasma formation, 8; release of, by recombination, 11; and response to selection, 11–13; in populations with and without sex, 13–16

variance, genetic: in fitness (additive), 170, 171, 172; in frequency of recombination, 9, 72, 73–7; in sex ratio, 148–50
variation, selectively neutral, 24
Volvocidae: size of colony, and degree of anisogamy in, 155

weevils, apomictic, 45–6; geographical distribution of, 54

Xyleborus (bark beetles), arrhenotoky in, 53